T0073183

METRIC SPACES AND RELATED ANALYSIS

Subiman Kundu
(retired) Indian Institute of Technology Delhi, India

Manisha Aggarwal
St. Stephen's College, University of Delhi, India

World Scientific

NEW JERSEY · LONDON · SINGAPORE · BEIJING · SHANGHAI · HONG KONG · TAIPEI · CHENNAI · TOKYO

Published by

World Scientific Publishing Co. Pte. Ltd.

5 Toh Tuck Link, Singapore 596224

USA office: 27 Warren Street, Suite 401-402, Hackensack, NJ 07601

UK office: 57 Shelton Street, Covent Garden, London WC2H 9HE

Library of Congress Control Number: 2023035290

British Library Cataloguing-in-Publication Data
A catalogue record for this book is available from the British Library.

METRICS SPACES AND RELATED ANALYSIS

Copyright © 2024 by World Scientific Publishing Co. Pte. Ltd.

All rights reserved. This book, or parts thereof, may not be reproduced in any form or by any means, electronic or mechanical, including photocopying, recording or any information storage and retrieval system now known or to be invented, without written permission from the publisher.

For photocopying of material in this volume, please pay a copying fee through the Copyright Clearance Center, Inc., 222 Rosewood Drive, Danvers, MA 01923, USA. In this case permission to photocopy is not required from the publisher.

ISBN 978-981-127-891-4 (hardcover)
ISBN 978-981-127-892-1 (ebook for institutions)
ISBN 978-981-127-893-8 (ebook for individuals)

For any available supplementary material, please visit
https://www.worldscientific.com/worldscibooks/10.1142/13486#t=suppl

Desk Editors: Nambirajan Karuppiah/Rok Ting Tan

Typeset by Stallion Press
Email: enquiries@stallionpress.com

To my sister Dr. Susmita Kundu who has always
encouraged me to write reader-friendly
good books in mathematics.

— Subiman Kundu

To my parents Satish Aggarwal and Poonam Aggarwal
who have taught me to believe in myself.

— Manisha Aggarwal

Foreword

Metric Spaces and Related Analysis is an introductory textbook on metric spaces aimed at advanced undergraduates and first-year graduate students. It presupposes a modest knowledge of elementary set theory as well as analysis on the real line. A course in which this text would be used prepares students for a subsequent course in general topology as well as courses in higher analysis, e.g. functional analysis or measure and integration. It is important to note that the authors give primacy to sequences over topology, making the material more accessible and appropriate for a student headed for applied analysis who might be less interested in abstract mathematics for its own sake.

What are the main topics to be covered in such a course? There is wide agreement on that: distance functions, balls, bounded and totally bounded sets, special points relative to a subset of a metric space (point of closure, accumulation point, interior point, boundary point), convergence of sequences in a metric space, continuity and uniform continuity, equivalent metrics, normed linear spaces as metric spaces, Cauchy sequences, completeness, compactness via sequences and open covers, pointwise and uniform convergence of sequences of functions, and connectedness notions. The authors handle all of these in a gentle, conversational way. The authors also include some standard but optional topics: Lipschitz functions, Baire metric spaces, function oscillation, locally compact metric spaces, and the Stone–Weierstrass theorem. Kundu and Aggarwal are intent on keeping their presentation as concrete as possible by frequently working in the usual contexts that analysts encounter, e.g. focusing on standard sequence spaces and the space $C[a, b]$ equipped with the supremum norm or L^p-distance.

But the authors include notable material that is simply not to be found in competitive texts. They pay considerable attention to the class of functions that map Cauchy sequences to Cauchy sequences that lies between the uniformly continuous functions and the continuous functions. They introduce two important classes of metric spaces lying between the compact metric spaces and the complete metric spaces: the UC spaces and the larger class of cofinally complete metric spaces. Both of these classes did not become generally known until after 1950. UC spaces are those on which each continuous function with values in a second metric space is automatically uniformly continuous. Cofinally complete metric spaces are those in which each cofinally Cauchy sequence has a convergent subsequence (a sequence (x_n) in a metric space (X, d) is called cofinally Cauchy provided for each positive ε there exists an infinite set of indices $\mathbb{N}_0$ such that whenever $n, k \in \mathbb{N}_0$, we have $d(x_n, x_k) < \varepsilon$.)

Kundu and Aggarwal also introduce an important family of subsets of a metric space lying between the totally bounded subsets and the full family of bounded subsets, called the finitely chainable subsets. This family of subsets is intrinsic to uniform continuity, and arises in a characterization of those metric spaces on which the real-valued uniformly continuous functions form a ring that was only discovered in the last 10 years by Javier Cabello Sánchez.

The inclusion of this novel material hopefully will convince the student and instructor alike that the study of metric spaces — particularly analysis on metric spaces — remains a vibrant field, and that there are notions of truly general interest that continue to be discovered.

Finally, the student should appreciate the authors' historical references as well as ample hints for the solution of exercises.

Gerald Beer
Burbank, California

Preface

This book offers a comprehensive study of one of the foundational topics in mathematics, known as metric spaces. The significance of abstract concepts and some nice notions related to metric spaces inspired us to write this book. Although there are many undergraduate texts for metric spaces, but the introduction of the concepts of UC spaces, cofinal completeness, and that of finite chainability makes the text unique of its kind. This will help the readers in

(i) taking the secondary step towards analysis on metric spaces.
(ii) realizing the connection between the two most important classes of functions, continuous functions and uniformly continuous functions.
(iii) understanding the gap between compact metric spaces and complete metric spaces.

Moreover, the reader will also find interesting exercises on various subtleties of continuity like subcontinuity, upper semi-continuity, lower semi-continuity, etc. This will motivate them to explore the special classes of functions to further extent. Additionally, special emphasis has been laid on Lipschitz functions. All these notions could be taken as topics for projects at undergraduate or graduate level. We have also referred to some research papers and textbooks in the bibliography towards the end. This will facilitate the detailed study of the concepts for the interested readers.

Certain sections have been concluded by the historical perspective of the significant results (given in the section). The introductory book covers a wide range of concepts in a very natural and concise manner. This makes the text suitable for not just the undergraduate students, but also for the graduate students as well as for the instructors. We expect the reader to have some basic knowledge of the set theory and real analysis for better understanding. Some preliminaries are also included in Sections 1.1 and 1.2 for making this book self-explanatory.

We feel greatly obliged that the anonymous reviewers put considerable effort and time to read the whole manuscript. Consequently, the detailed reports by the reviewers helped in substantial additions across the length and breadth of the book. Finally, we would like to express our gratitude to Prof. Gerald Beer who has written the foreword for this book.

Contents

Chapter 1

Fundamentals of Analysis

We begin with the introduction of the fundamental structure of metric spaces along with various useful illustrations. Since the foundation of metric spaces lies in real analysis, we also discuss some basic relevant notions of real analysis along with that of set theory which are needed in the upcoming chapters.

1.1 Some Elements of Set Theory

1.1.1 *De Morgan's laws*

First recall that if X is a set and $A \subseteq X$, then A^c denotes the complement of A in X which is given by

$$A^c = X \setminus A = \{x \in X : x \notin A\}.$$

Theorem 1.1.1 (De Morgan's laws). *If X is a set and $\{A_i : i \in I\}$ is a family of subsets of X, then*

(a) $\left(\bigcap_{i \in I} A_i \right)^c = \bigcup_{i \in I} (A_i)^c,$

(b) $\left(\bigcup_{i \in I} A_i \right)^c = \bigcap_{i \in I} (A_i)^c.$

1

Proof. We prove (a), then (b) could be proved in a similar way.

$$x \in \left(\bigcap_{i \in I} A_i \right)^c$$

$$\Leftrightarrow x \notin \bigcap_{i \in I} A_i$$

$$\Leftrightarrow x \notin A_i \quad \text{for some } i \in I$$

$$\Leftrightarrow x \in (A_i)^c \quad \text{for some } i \in I$$

$$\Leftrightarrow x \in \bigcup_{i \in I} (A_i)^c.$$

$\square$

Thus, union and intersection of sets are dual concepts with respect to complementation.

Consider a function $f : X \to Y$ between two sets X and Y. For a subset A of X, the image $f(A)$ of A under f is defined by: $f(A) = \{f(a) : a \in A\}$. Analogously, the inverse image of a subset B of Y under f is given by: $f^{-1}(B) = \{x \in X : f(x) \in B\}$. It is evident that $f(A) \subseteq Y$, while $f^{-1}(B) \subseteq X$. Later on, especially in Chapter 2, we shall be using the following relations: if $\{A_i : i \in I\}$ and $\{B_i : i \in I\}$ are families of subsets of X and Y respectively then we have

(i) $f(\bigcup_{i \in I} A_i) = \bigcup_{i \in I} f(A_i)$

(ii) $f(\bigcap_{i \in I} A_i) \subseteq \bigcap_{i \in I} f(A_i)$

(iii) $f^{-1}(\bigcup_{i \in I} B_i) = \bigcup_{i \in I} f^{-1}(B_i)$

(iv) $f^{-1}(\bigcap_{i \in I} B_i) = \bigcap_{i \in I} f^{-1}(B_i)$

(v) $f^{-1}(B^c) = (f^{-1}(B))^c.$

1.1.2 *Equivalence relation*

Definition 1.1.1. Let X be a non-empty set. A subset of $X \times X$ is called a *relation* on X, usually denoted by R.

Since X is taken twice, that is, $X \times X$, such a relation is more precisely a *binary* relation. But we will simply use the term, 'relation'.

If $(x, y) \in R$, we also write, xRy. There can be many types of relations. But among these, two are most important: Equivalence and Partial order. In order to define these two, we need the following definitions first.

Definitions 1.1.2. (a) *Reflexivity:* If xRx for all $x \in X$, that is, if $(x, x) \in R$ for all $x \in X$, then R is called **reflexive**.
(b) *Symmetry:* Whenever xRy, then yRx for x, $y \in X$.
(c) *Antisymmetry:* If xRy and yRx then $x = y$.
(d) *Transitivity:* If xRy and yRz then xRz. Note that xRz happens through the transit of y.

Now if a relation R satisfies (a), (b) and (d), then it is called an **equivalence relation**. So an equivalence relation is reflexive, symmetric and transitive. An equivalence relation is usually denoted by $\sim$ or $\approx$. On the other hand, if a relation R satisfies (a), (c) and (d), then it is called a **partial order**. So a partial order is reflexive, anti-symmetric and transitive. Usually a partial order is denoted by $\leq$.

Example 1.1.1. Let $\mathcal{F}$ be the collection of all finite subsets of the set of natural numbers $\mathbb{N}$. For A, $B \in \mathcal{F}$, define $A \sim B$ if A and B have the same number of elements. Evidently, $\sim$ is an equivalence relation on $\mathcal{F}$.

Example 1.1.2. Consider the set $\mathbb{Z} \times (\mathbb{Z} \setminus \{0\})$, where $\mathbb{Z}$ denotes the set of integers. Let (m, n), $(p, q) \in \mathbb{Z} \times (\mathbb{Z} \setminus \{0\})$, define

$$(m, n) \sim (p, q) \quad \text{if} \quad mq = np.$$

The reader can easily verify that $\sim$ is indeed an equivalence relation on $\mathbb{Z} \times (\mathbb{Z} \setminus \{0\})$.

Example 1.1.3. On the set $\mathbb{R}$ of all real numbers, define the usual relation between the numbers: $x \leq y$, x, $y \in \mathbb{R}$. Clearly, $\leq$ is a partial order.

Example 1.1.4. Let X be a set and $\mathcal{P}(X)$ be the set of all subsets of X. Then the set inclusion relation $\subseteq$ is a partial order on the power set $\mathcal{P}(X)$.

Coming back to equivalence relation. Let $\sim$ be an equivalence relation on a (non-empty) set X. Let $x \in X$ and define $[x] = \{y \in$

$X : y \sim x\}$. Such a set is called the **equivalence class** of x with respect to $\sim$. Note that $[x]$ is a subset of X containing x (why?). Also note that for all x, $y \in X$, either $[x] = [y]$ or else $[x] \cap [y] = \emptyset$ (prove it!). So this fact gives rise to a partition of X into pairwise disjoint sets. But you can also prove that a partition of X gives an equivalence relation. So an equivalence relation on X corresponds to a partition of X and vice-versa. In Example 1.1.2, the equivalence classes with respect to $\sim$ are nothing but the rational numbers.

1.1.3 *Cartesian product and axiom of choice*

If $\{X_1, X_2, \ldots, X_n\}$ is a family of (finitely many) non-empty sets, then the Cartesian product of this family is defined to be

$$\prod_{i=1}^{n} X_i = X_1 \times X_2 \times \cdots \times X_n$$

$$= \{(x_1, x_2, \ldots, x_n) : x_i \in X_i \quad \text{for } 1 \leq i \leq n\}.$$

Note that in order to define this product, we have used an order of the sets $\{X_i : 1 \leq i \leq n\}$. But if we have an uncountably infinite family of sets, we do not have such an order. Suppose for each $\alpha \in \mathbb{R}$, we have a non-empty set X_α. Then *how do we define the Cartesian product of the family $\{X_\alpha : \alpha \in \mathbb{R}\}$?* So in order to define the Cartesian product (hereafter simply called 'product') of an arbitrary family of sets, we need to get rid of the order mentioned earlier. Thus, we need to redefine the product $\prod_{i=1}^{n} X_i$ of finitely many sets $X_1, X_2, \ldots, X_n$ in terms of functions. Closely look at the definition of a function- it does not involve an order.

Given a finite family of non-empty sets $X_1, X_2, \ldots, X_n$, define a function $f : \{1, 2, \ldots, n\} \to \bigcup_{i=1}^{n} X_i$ such that $f(i) \in X_i$. If we denote $f(i)$ by x_i, then since the domain of f is $\{1, 2, \ldots, n\}$, a finite set, we can describe f simply by listing its values $(x_1, x_2, \ldots, x_n)$. This way the element $(x_1, x_2, \ldots, x_n)$ of $X_1 \times X_2 \times \ldots \times X_n$ can be identified with the function f. Note two things about this function f:

(a) The domain of f is $\{1, 2, \ldots, n\}$ which is the index set of the family of sets $\{X_1, X_2, \ldots, X_n\}$. This index set is used to identify the sets in the family of the sets. Here 1 is used to identify the set X_1, while in general, i $(1 \leq i \leq n)$ is used to identify the set

X_i. So there has to be a one-to-one correspondence between an index set I and the corresponding family of sets $\{X_i : i \in I\}$.

(b) The range of the function f is $\bigcup_{i=1}^{n} X_i$, but there is a restriction (or choice) on f: $f(i) \in X_i$. Actually such a function is called a *choice function*.

Now we can define the product of a family of non-empty sets, $\{X_i : i \in I\}$, as follows:

$$\prod_{i \in I} X_i = \{f : I \to \bigcup_{i \in I} X_i : f(i) \in X_i \ \forall \ i \in I\}.$$

Thus, $\prod_{i \in I} X_i$ is the collection of choice functions. But, given an arbitrary family $\{X_i : i \in I\}$ of non-empty sets, what is the guarantee that we will have such a choice function? One may say that we will just keep picking up one from each X_i. But how long will you keep picking up? For the pick-up process to be valid, we need to ensure that it is completed in finite time. Think about the use of the *Principle of Mathematical Induction*!

Let us reformulate the question about the choice function.

Given a non-empty family $\{X_i : i \in I\}$ of non-empty sets, is the Cartesian product $\prod_{i \in I} X_i$ non-empty?

In other words, $I \neq \emptyset$, each $X_i \neq \emptyset \Rightarrow \prod_{i \in I} X_i \neq \emptyset$? Though the question appears to be simple, the answer does not appear so. In fact, though the answer is expected to be affirmative, we can't prove it. The affirmative answer to this question is known as the *Axiom of Choice*: If $\{X_i : i \in I\}$ is a non-empty family of sets such that X_i is non-empty for each $i \in I$, then $\prod_{i \in I} X_i$ is non-empty.

A useful equivalent formulation of the axiom of choice is stated below which can be easily verified by the readers:

Proposition 1.1.1. *The following statements are equivalent:*

(a) *If $\{X_i : i \in I\}$ is a non-empty family of sets such that X_i is non-empty for each $i \in I$, then $\prod_{i \in I} X_i$ is non-empty.*

(b) *If $\{X_i : i \in I\}$ is a non-empty family of pairwise disjoint sets such that each X_i is non-empty, then there exists a non-empty subset E of $\bigcup_{i \in I} X_i$ such that $E \cap X_i$ consists of precisely one element for each $i \in I$.*

Remarks. (i) Here 'pairwise disjoint' means $X_i \cap X_j = \emptyset$ for any distinct pair $\{i, j\}$ from I, that is, $i, j \in I$ and $i \neq j$.

(ii) Note that (a) gives the existence of a choice function, while (b) gives the existence of a choice set.

(iii) Any one of the equivalent statements given in the previous proposition is known as the Axiom of Choice.

(iv) There are many more equivalent formulations of the axiom of choice, but their proofs are not usually easy. But one such important equivalent formulation is Zorn's Lemma.

In this course, we will encounter several applications of the axiom of choice. In many proofs, we may not even realize the application of the axiom.

1.2   The Real Numbers

The objective of this section is to study some crucial properties of the set of real numbers $\mathbb{R}$ which are needed in the upcoming text. The first one is the completeness axiom or the axiom of continuity. In order to discuss it, we need to talk about various kinds of 'bounded' subsets of $\mathbb{R}$.

Definitions 1.2.1. Let A be a non-empty subset of $\mathbb{R}$. Then A is called **bounded above** if there exists $b \in \mathbb{R}$ such that $a \leq b$ holds for all a in A and b is called an **upper bound** of A.

Note that if b is an upper bound of A and if $b' > b$, $b' \in \mathbb{R}$, then b' is also an upper bound of A. Moreover, the set $S = \{b \in \mathbb{R} : b$ is an upper bound of $A\}$ is non-empty for the bounded above set A. Now the question arises: *Does S have a least element ? Can you prove it?*

Definition 1.2.2. Let A be a non-empty subset of $\mathbb{R}$. Suppose A is bounded above. If the set $S = \{b \in \mathbb{R} : b$ is an upper bound of $A\}$ has a least element s (say), then s is called a **least upper bound** (lub) or **supremum** of A.

Evidently, if S has a least element, then it must be unique. So it should be referred to as *the* least element or *the* supremum. But the definition of supremum does not answer our query: Does A have a supremum? Before answering this query, we would like to deal with the symmetric problem of the subsets of $\mathbb{R}$ which are bounded below.

Definitions 1.2.3. Let A be a non-empty subset of $\mathbb{R}$. Then A is called **bounded below** if there exists $b \in \mathbb{R}$ such that $b \leq a$ holds for all a in A and b is called a **lower bound** of A.

Again note that if b is a lower bound of A and if $b' < b$, $b' \in \mathbb{R}$, then b' is also a lower bound of A. So in this case we have the following query: Does the non-empty set $S = \{b \in \mathbb{R} : b$ is a lower bound of $A\}$ have a greatest element ? If yes, then such an element is called *the* **greatest lower bound** (glb) or *the* **infimum** of A. Now observe the following:

Proposition 1.2.1. *The following statements are equivalent:*

(a) *Every non-empty subset A of $\mathbb{R}$, which is bounded above, has a supremum.*
(b) *Every non-empty subset A of $\mathbb{R}$, which is bounded below, has an infimum.*

Proof. It can be proved by observing that if $\emptyset \neq A \subseteq \mathbb{R}$, then A is bounded above if and only if $A' = \{-a : a \in A\}$ is bounded below. $\qquad\square$

Note that in the previous proposition, we only prove that (a) $\Leftrightarrow$ (b), that is, if one of them is true, the other one is also true. We do not prove that either (a) is true or (b) is true. So we need to assume one of those two equivalent conditions as an axiom. It is well known as the Completeness axiom on $\mathbb{R}$ or the Axiom of Continuity on $\mathbb{R}$.

Completeness axiom on $\mathbb{R}$: *Every non-empty set of real numbers that is bounded above has a least upper bound.*

It should be noted that the lub need not belong to the set. If it belongs to the set, we call it the *maximum* of the set.

Example 1.2.1. Consider the sets $A = [0,1)$ and $B = [0,1]$. For both the sets, the supremum is 1. While $1 \in B$, $1 \notin A$. Hence we can say that 1 is the maximum element of B, while A does not have a maximum.

The following result is useful in the analysis of metric spaces. Their proofs are left as an exercise for the interested readers.

Theorem 1.2.1. *Let A be a non-empty subset of $\mathbb{R}$ such that A is bounded above. Then given $\epsilon > 0$, there exists $x_\epsilon \in A$ (x_ϵ depends on ϵ) such that $\sup(A) - \epsilon < x_\epsilon \leq \sup(A)$. Here $\sup(A)$ denotes the supremum of A.*

Remark. In the previous theorem, the expression '$x_\epsilon \leq \sup A$' may not be replaced by '$x_\epsilon < \sup A$'. For example, look at the set $A = [0,1] \cup \{2\}$. Here $\sup A = 2$. Can you find $x_\epsilon \in A$ such that $x_\epsilon \neq \sup A$ and $2 - \frac{1}{2} = \frac{3}{2} < x_\epsilon$?

Corollary 1.2.1. *The set $\mathbb{N}$ of all natural numbers is not bounded above.*

The glb (infimum) counterpart of Theorem 1.2.1 is as follows:

Theorem 1.2.1. *Let A be a non-empty subset of $\mathbb{R}$ such that A is bounded below. Then given $\epsilon > 0$, there exists $x_\epsilon \in A$ (x_ϵ depends on ϵ) such that $\inf(A) \leq x_\epsilon < \inf(A) + \epsilon$. Here $\inf A$ denotes the infimum of A.*

Now we discuss briefly one of the useful properties of real numbers, namely the 'Archimedean Property'. Instead of the 'Archimedean Property', it should have been called the 'Eudoxusian Property' in honour of Eudoxus. The property was used extensively by the Greek mathematician Archimedes, but most probably, it was introduced by the Greek scholar Eudoxus.

Theorem 1.2.2 (The Archimedean property). *If x and y are two positive real numbers, then there exists some natural number n such that $nx > y$.*

Proof. Suppose, if possible, $nx \leq y$ for all $n \in \mathbb{N}$. Then $n \leq y/x \; \forall \, n \in \mathbb{N}$, as $x > 0$. But this implies that the set of all natural numbers $\mathbb{N}$ is bounded above by a fixed real number y/x. This is impossible. $\qquad\qquad\qquad\qquad\qquad\qquad\qquad\qquad\qquad\qquad\square$

Note that the importance of Theorem 1.2.2 lies in the following fact: You choose any two positive reals x and y (x may be very small, while y may be very large), but we can find a sufficiently large n in $\mathbb{N}$ such that we can increase the length of x by multiplying it by n so that nx becomes larger than y.

1.3 Metric Spaces

Now we are ready to move towards the abstract study of metric spaces. But before explaining a metric d and the corresponding metric space, we should have a closer look at the usual distance on $\mathbb{R}$ given by $|\cdot|$. Note that if x and y are two real numbers, then the distance from x to y is given by $|x - y|$, while the distance from y to x is given by $|y - x|$. But they are same. Hence it is instead referred to as the distance between x and y. While doing some manipulations with sequences in $\mathbb{R}$, often we use the triangle inequality:

$$|x - y| \leq |x - z| + |z - y| \ \forall \ x, \ y, \ z \in \mathbb{R}.$$

Other two important facts about this distance are so natural that often we forget to notice them: (i) the distance between x and y is always non-negative, that is, $|x - y| \geq 0$, and $|x - y| = 0 \Leftrightarrow x = y$, (ii) $|x - y| = |y - x|$. You may go through your knowledge of calculus and analysis on $\mathbb{R}$. You will be surprised to discover that the primary tools in $\mathbb{R}$ are given by (i) and (ii) and the triangle inequality (also denoted by $\triangle-$ inequality). In order to do similar analysis on various other sets conveniently, there was a need to generalize this idea of distance on an arbitrary non-empty set X. For this purpose, a metric (or a distance) on X is defined as follows.

Definition 1.3.1. A **metric** (or a distance) d on a non-empty set X is a function $d : X \times X \to \mathbb{R}$ satisfying the following properties: for every $x, \ y, \ z \in X$ we have

(a) $d(x, y) \geq 0$ and $d(x, y) = 0 \Leftrightarrow x = y$;
(b) $d(x, y) = d(y, x)$ *(symmetry)*;
(c) $d(x, y) \leq d(x, z) + d(z, y)$ *(the triangle inequality)*.

Furthermore, the pair (X, d) is called a **metric space**.

Most of the times, the reference to the metric d in the notation (X, d) is omitted in case there is no ambiguity. An immediate application of the triangle inequality is to prove the following basic result:

Proposition 1.3.1. *Let (X, d) be a metric space. Then*

$$|d(x, z) - d(y, z)| \leq d(x, y) \ \forall \ x, \ y, \ z \in X.$$

Proof. By the triangle inequality, we have $d(x, z) \le d(x, y) + d(y, z)$ and hence $d(x, z) - d(y, z) \le d(x, y)$. Now by interchanging the roles of x and y, we get $d(y, z) - d(x, z) \le d(y, x) = d(x, y)$. Hence, $|d(x, z) - d(y, z)| \le d(x, y)$.   □

Now let us look at some standard but useful examples of metric spaces. We start with the example which was the main motivation behind this concept.

Example 1.3.1. Consider the set of all real numbers $\mathbb{R}$ with the usual distance $d(x, y) = |x - y| \quad \forall \, x, \, y \in \mathbb{R}$. Then $(\mathbb{R}, d)$ is a metric space. Note that the metric d is well known as the usual (or standard) metric on $\mathbb{R}$.

Example 1.3.2. Consider $\mathbb{R}^n$ which is the set of all ordered n-tuples $x = (x_1, \ldots, x_n)$ of real numbers. In other words, $\mathbb{R}^n = \prod_{i=1}^{n} X_i$, where $X_i = \mathbb{R}$ for all $i \in \{1, 2, \ldots, n\}$. Define a function d on $\mathbb{R}^n \times \mathbb{R}^n$ as follows:

$$d(x, y) = \left(\sum_{i=1}^{n} (x_i - y_i)^2 \right)^{1/2}$$

for $x = (x_1, \ldots, x_n)$ and $y = (y_1, \ldots, y_n)$ in $\mathbb{R}^n$. Then the function d satisfies the properties of a metric: the triangle inequality can be proved using Minkowski's inequality (see Appendix), while the verification of the rest of the properties is straightforward. The metric d is called the **Euclidean metric** and $\mathbb{R}^n$ equipped with this metric is referred to as *n-dimensional Euclidean space*. Note that this Euclidean distance on $\mathbb{R}^2$ and $\mathbb{R}^3$ actually gives the Euclidean geometry.

Note. *When $n = 1$, then the Euclidean metric is nothing but the usual metric on $\mathbb{R}$. Unless mentioned explicitly, $\mathbb{R}^n$ and its subsets are assumed to be equipped with the Euclidean metric in this book.*

Example 1.3.3. One needs to be careful while defining a metric on the set of all ordered n-tuples of complex numbers, $\mathbb{C}^n$, which is analogous to the Euclidean metric on $\mathbb{R}^n$. If $z = (z_1, \ldots, z_n)$ and $w = (w_1, \ldots, w_n) \in \mathbb{C}^n$, then define

$$d(x, y) = \left(\sum_{i=1}^{n} |z_i - w_i|^2 \right)^{1/2}$$

(here we need to write $|z_i - w_i|^2$, instead of $(z_i - w_i)^2$, in order to ensure that $|z - w|$ is a real number.) Note that $\mathbb{C}^n$ equipped with this metric d is sometimes called *complex Euclidean n-space* or *n-dimensional unitary space*.

Note that we can have multiple metrics defined on a set with at least two points. So using next example, we can construct more metrics on $\mathbb{R}^n$.

Example 1.3.4. Let $\{(X_i, d_i) : 1 \leq i \leq n\}$ be a finite collection of metric spaces. Then the reader should verify that the following three functions define metrics on the Cartesian product $\prod_{i=1}^n X_i = \{(x_1, x_2, \ldots, x_n) : x_i \in X_i$ for $1 \leq i \leq n\}$:

(i) $\mu_1(a, b) = \sum_{i=1}^n d_i(a_i, b_i)$,

(ii) $\mu_2(a, b) = \sqrt{\sum_{i=1}^n (d_i(a_i, b_i))^2}$,

(iii) $\mu_\infty(a, b) = \max\{d_i(a_i, b_i) : 1 \leq i \leq n\}$,

where $a = (a_1, \ldots, a_n)$, $b = (b_1, \ldots, b_n) \in \prod_{i=1}^n X_i$. Moreover, we have

$$\mu_\infty(a, b) \leq \mu_2(a, b) \leq \mu_1(a, b) \leq n\mu_\infty(a, b).$$

Example 1.3.5. An interesting metric can be defined on every nonempty set X in a trivial way. For $x \neq y$, define $d(x, y) = 1$ and set $d(x, x) = 0$. The metric d is called the **discrete metric** or **trivial metric** on X.

The discrete metric given in Example 1.3.5 is often used as an easy counter-example (will see later). To avoid confusion in the remaining text, we would like to explicitly define discrete sets.

Definition 1.3.2. A subset A of a metric space (X, d) is called **discrete** if $\forall x \in A \; \exists \delta_x > 0$ such that $d(x, y) \geq \delta_x \; \forall y \in A \setminus \{x\}$.

Example 1.3.6. Let X be a nonempty set and (Y, ρ) be a metric space. Consider a one-to-one function $f : X \to (Y, \rho)$. Then using f, we can define a metric d on X as follows: $d(x, x') = \rho(f(x), f(x'))$ for $x, x' \in X$.

Note that if f is not one-to-one, then d is called pseudometric because $d(x, x') = 0$ need not imply $x = x'$ for $x, x' \in X$. More precisely, a function $d : X \times X \to \mathbb{R}$ is called a **pseudometric** if $d(x,y) \geq 0$; $d(x,x) = 0$; $d(x,y) = d(y,x)$; and $d(x,y) \leq d(x,z) + d(z,y)$ for all x, y, $z \in X$.

Interestingly, there is a standard technique of inducing a metric space from a pseudometric space: suppose (X,d) is a pseudometric space. Define an equivalence relation (that is, reflexive, symmetric and transitive) on X by: $x \sim z$ if and only if $d(x,z) = 0$. Now let $S = \{E_x : x \in X\}$, where E_x denotes the equivalence class containing x, and define a function δ on $S \times S$ as: $\delta(E_x, E_z) = d(x,z)$. It can be easily verified that δ defines a well-defined metric on $S \times S$.

Let P_n be the set of all real polynomials with degree less than or equal to n. Define a function:

$$f : P_n \to \mathbb{R}^{n+1}$$

$$f(a_o + a_1 x + \cdots + a_n x^n) = (a_o, a_1, \ldots, a_n).$$

Then note that f is an injective function and hence by using Example 1.3.6, we can get metrics on P_n. *Can you think of a metric on the set of all real polynomials?*

Example 1.3.7. Consider ℓ^p ($p \geq 1$ is a fixed real number), the set of all sequences (x_n) in $\mathbb{R}$ (we can also consider sequences of complex numbers) such that $\sum_{k=1}^{\infty} |x_k|^p < \infty$, equipped with the metric

$$d(x,y) = \left(\sum_{k=1}^{\infty} |x_k - y_k|^p \right)^{1/p},$$

where $x = (x_k)$ and $y = (y_k)$. Note that from Minkowski's inequality (see Appendix) it follows that (ℓ^p, d) forms a metric space.

Remark. If $p < 1$, then the triangle inequality fails: let a be the zero sequence, $b = (1,1,0,0,\ldots)$ and $c = (0,1,0,0,\ldots)$. Then $d(a,c) = 1 = d(c,b)$ and $d(a,b) = 2^{1/p} > d(a,c) + d(c,b)$ because $p < 1$.

Example 1.3.8. Consider ℓ^∞, the set of all bounded sequences (x_n) (that is, $\sup_{n \in \mathbb{N}} |x_n| < \infty$). Let us define a function,

$$d(x,y) = \sup_{n \in \mathbb{N}} |x_n - y_n|$$

for $x = (x_n) \in \ell^\infty$ and $y = (y_n) \in \ell^\infty$. We only need to look at the proof for the triangle inequality: for each $n \in \mathbb{N}$ we have,

$$|x_n - y_n| \le |x_n - z_n| + |z_n - y_n| \le \sup_{n \in \mathbb{N}} |x_n - z_n| + \sup_{n \in \mathbb{N}} |z_n - y_n|.$$

Thus, $d(x,y) \le d(x,z) + d(z,y)$.

In Example 1.3.8, we considered the set of all bounded sequences so that the supremum metric is well-defined. But what if we need to define a metric on the set of all real sequences (that is, sequences in $\mathbb{R}$)? In fact, we try to construct a metric on a general set in the next example.

Example 1.3.9. Let I be an arbitrary indexing set. Consider $\mathbb{R}^I = \prod_{i \in I} \mathbb{R} = \{(x_i)_{i \in I} : x_i \in \mathbb{R}\}$. (Note that the elements of $\mathbb{R}^I$ are nothing but functions from I to $\mathbb{R}$: $i \mapsto x_i$.) Then the following function defines a metric on $\mathbb{R}^I$,

$$d(x,y) = \sup\{\rho(x_i, y_i) : i \in I\}$$

for $x = (x_i)_{i \in I}$ and $y = (y_i)_{i \in I}$ in $\mathbb{R}^I$. Here $\rho(x_i, y_i) = \min\{|x_i - y_i|, 1\}$ is a bounded metric on $\mathbb{R}$ (refer to Exercise 1.3.7). The metric d is called the **uniform metric** on $\mathbb{R}^I$.

Now can you think of a metric on the set of all real polynomials?

Example 1.3.10. Let $\{(X_n, d_n) : n \in \mathbb{N}\}$ be a collection of metric spaces. Then

$$d(x,y) = \sum_{n=1}^\infty \frac{1}{2^n} \frac{d_n(x_n, y_n)}{1 + d_n(x_n, y_n)},$$

where $x = (x_n)$ and $y = (y_n)$, defines a metric on the Cartesian product $\prod_{i=1}^\infty X_i$: for the triangle inequality, first note that the function $\delta_n(x,y) = \frac{d_n(x,y)}{1 + d_n(x,y)}$ defines a metric on X_n for $n \in \mathbb{N}$ (see Exercise 1.3.7). Thus,

$$\delta_n(x_n, y_n) \le \delta_n(x_n, z_n) + \delta_n(z_n, y_n) \text{ for } x_n, y_n, z_n \in X_n.$$

This implies that

$$\sum_{n=1}^{k} \frac{1}{2^n} \delta_n(x_n, y_n) \leq \sum_{n=1}^{k} \frac{1}{2^n} \delta_n(x_n, z_n) + \sum_{n=1}^{k} \frac{1}{2^n} \delta_n(z_n, y_n)$$

$$\leq \sum_{n=1}^{\infty} \frac{1}{2^n} \delta_n(x_n, z_n) + \sum_{n=1}^{\infty} \frac{1}{2^n} \delta_n(z_n, y_n)$$

$$= d(x, z) + d(z, y) \quad \text{for each } k \in \mathbb{N}.$$

Hence $d(x, y) \leq d(x, z) + d(z, y)$ for $x = (x_n)$, $y = (y_n)$, $z = (z_n) \in \prod_{i=1}^{\infty} X_i$.

Example 1.3.11. Consider the set $C[a, b]$ of all continuous real-valued functions on a closed and bounded interval $[a, b]$. (Recall that a function $f : ([a, b], |\cdot|) \to (\mathbb{R}, |\cdot|)$ is said to be *continuous* at $x_o \in [a, b]$ if given $\epsilon > 0$, there exists a $\delta > 0$ (δ depends on ϵ and x_o) such that $|f(x) - f(x_o)| < \epsilon$ whenever $|x - x_o| < \delta$ and $x \in [a, b]$.) Then the function D defined as

$$D(x, y) = \max_{t \in [a,b]} |x(t) - y(t)|,$$

where x, y are continuous functions on $[a, b]$, is a metric on $C[a, b]$. The metric D is called the **uniform metric**. Here one only needs to verify the triangle inequality as the verification of the rest of the properties is straightforward. Now for all $t \in [a, b]$, we have

$$|x(t) - y(t)| \leq |x(t) - z(t)| + |z(t) - y(t)| \leq \max_{t \in [a,b]} |x(t) - z(t)|$$

$$+ \max_{t \in [a,b]} |z(t) - y(t)|.$$

Thus, $D(x, y) \leq D(x, z) + D(z, y)$ for all x, y, $z \in C[a, b]$.

Since every real-valued continuous function on a closed and bounded interval $[a, b]$ attains its bounds, it is justified to use 'maximum' in the definition of the metric d in the previous example. But we cannot use the same definition for the collection of all real-valued bounded functions on $[a, b]$ (will see in Example 1.3.13). Further, note that geometrically this metric d denotes the largest vertical distance between the graphs of the functions x and y.

Example 1.3.12. The set $C[a,b]$ can also be equipped with the following metric:

$$d_p(x,y) = \left(\int_a^b |x(t) - y(t)|^p dt \right)^{1/p},$$

where $1 \le p < \infty$.

Clearly, $d_p(x,x) = 0$ and $d_p(x,y) \ge 0$ for all x, $y \in C[a,b]$. Now for proving that $d_p(x,y) = 0 \Rightarrow x = y$, one needs to observe the following:

Claim. If $f \in C[a,b]$ with $f(t) \ge 0$ for all $t \in [a,b]$ and $\int_a^b f(t)dt = 0$, then f is identically 0 on $[a,b]$.

Suppose, if possible, $f(t_o) > 0$ for some $t_o \in [a,b]$. Then by the continuity of f at t_o, there exists a $\delta > 0$ such that

$$|f(t) - f(t_o)| < \frac{f(t_o)}{2} \text{ whenever } |t - t_o| < \delta \text{ and } t \in [a,b].$$

Consequently,

$$\int_a^b f(t)dt \ge \int_{t_o-\delta}^{t_o+\delta} f(t)dt \ge \int_{t_o-\delta}^{t_o+\delta} \frac{f(t_o)}{2}dt = f(t_o)\delta > 0.$$

This gives a contradiction. Hence $f(t) = 0$ for all $t \in [a,b]$. Thus,

$$d_p(x,y) = 0 \Rightarrow x = y \quad \text{for } x,\, y \in C[a,b].$$

The function d_p is clearly symmetric. We are only left with triangle's inequality which follows from Minkowski's inequality for integrals. Let us state the inequality without proof:

$$\left(\int_a^b |x(t) + y(t)|^p dt \right)^{1/p} \le \left(\int_a^b |x(t)|^p dt \right)^{1/p} + \left(\int_a^b |y(t)|^p dt \right)^{1/p},$$

for $1 \le p < \infty$ and $x,\, y \in C[a,b]$.

Remarks. (a) Geometrically, the metric d_1 represents the area between the graphs of the functions x and y.

(b) It should be noted that d_1 is a pseudometric on the set of all Riemann integrable functions on $[a, b]$. But it is not a metric on the set: consider the characteristic function:

$$x(t) = \begin{Bmatrix} 1 & \text{if } t = a \\ 0 & \text{if } a < t \le b \end{Bmatrix}$$

and $y(t) = 0$ for all $t \in [a, b]$. Then x and y are two distinct functions with $d_1(x, y) = 0$.

Example 1.3.13. Let X be a non-empty *set* and let $B(X)$ be the set of all real-valued bounded functions defined on X, that is,

$$B(X) = \{f : X \to \mathbb{R} : f(X) \text{ is bounded in } \mathbb{R}\}$$
$$= \{f : X \to \mathbb{R} : \exists \ M_f > 0 \text{ such that } |f(x)| \le M_f \ \forall \ x \in X\}.$$

On $B(X)$, define the metric D as follows:

$$D(f, g) = \sup\{|f(x) - g(x)| : x \in X\}.$$

This supremum is a non-negative real number since f and g are bounded functions. Moreover, the reader should check that D satisfies the properties of a metric on $B(X)$. The metric D is called the **supremum metric** or **uniform metric** on $B(X)$. The justification for calling it a 'uniform' metric will be given later.

Remark. (a) Note that even if we replace $\mathbb{R}$ by $\mathbb{C}$, that is, if we take $B(X)$ to be the set of all complex valued bounded functions on X, the function D defines a metric on this new collection as well.

(b) When $X = \mathbb{N} =$ the set of all natural numbers, $B(X)$ is nothing but l^∞. If we want to make a distinction between the set of all bounded real sequences and that of bounded complex sequences, we may use the notation $l^\infty(\mathbb{R})$ or $l^\infty(\mathbb{C})$. The l^∞-space plays a crucial role in measure theory and functional analysis.

(c) Since every continuous function on a closed and bounded interval is bounded, $C[a, b]$ is a subset of $B(X)$, where $X = [a, b]$.

(d) Note that Example 1.3.9 actually talks about the metric on the set of all real-valued functions defined on a set I. The elements of $\mathbb{R}^I$ can also be represented in the form of functions of the type $f : I \to \mathbb{R}$. So $f(i) = x_i$.

Towards the end of this section, we would like to precisely define bounded sets in a metric space. Of course, the definition should be consistent with the definition of bounded sets in $\mathbb{R}$. Recall that a subset A of $\mathbb{R}$ is *bounded* if there exists $M > 0$ such that $|x| \leq M$ for all $x \in A$, that is, $|x - 0| \leq M \; \forall \; x \in A$. But in a general metric space (X, d), we may not have the point zero. Observe that 'zero' actually works as a reference point from which we travel to other points in $\mathbb{R}$. In fact, we can choose any point x_o in a metric space (X, d) and can use it as a reference point in X. So we can define a subset A of a metric space (X, d) to be *bounded* if there exists $M > 0$ such that $d(x, x_o)$ (in place of $|x - 0|$) is less than M for all $x \in A$. But one may raise a question: how do we ensure that this definition is not dependent on the choice of x_o. This is ensured by the following result, the proof of which is left as an exercise for the reader.

Proposition 1.3.2. *Let (X, d) be a metric space, A be a non-empty subset of X and $x_o \in X$. Then the following are equivalent:*

(a) *there exists $M_1 > 0$ such that $d(x, x_o) \leq M_1$ for all $x \in A$.*
(b) *$\sup\{d(x, y) : x, \; y \in A\}$ is a real number, that is, there exists $M_2 > 0$ such that $d(x, y) \leq M_2$ for all $x, \; y \in A$.*

The authors use (b) of the last result as the definition of a bounded set in a metric space (X, d).

Definition 1.3.3. Let (X, d) be a metric space and A be a non-empty subset of X.

(a) The **diameter** of A, denoted by $d(A)$ or $\text{diam}(A)$, is defined as follows: $d(A) = \sup\{d(x, y) : x, \; y \in A\}$.
(b) If $d(A) \in \mathbb{R}$, then we say that A is **bounded** in (X, d).

The way in which diameter of a set is defined, completely justifies the terminology 'diameter' used for it.

Remark. (i) Since $d(x, y) \geq 0 \; \forall \; x, \; y \in X$, $d(A) \geq 0$. Often we write $d(A) < \infty$ in order to indicate that $d(A)$ is a real number.
(ii) A sequence (x_n) in a metric space (X, d) is called bounded if the set $A = \{x_n : n \in \mathbb{N}\}$ is bounded in (X, d).
(iii) A function $f : (X, d) \to (Y, \rho)$ is said to be bounded if the set $f(X)$ is bounded in (Y, ρ).

Evidently, in a metric space (X, d), every non-empty subset of a bounded set in (X, d) is again bounded. In particular, we have

Proposition 1.3.3. *Let (X, d) be a metric space and A be a non-empty subset of X. Then A is bounded in (X, d) if and only if every sequence in A is bounded.*

Exercises

1.3.1. Let d be a metric on a non-empty set X. Prove that the function, defined by $d'(x, y) = (d(x, y))^\alpha$ for x, $y \in X$ and fixed $\alpha \in (0, 1)$, is a metric on X. Is ρ also a metric on X, where $\rho(x, y) = (d(x, y))^2$ for x, $y \in X$?

1.3.2. If d_1 and d_2 are two metrics on a non-empty set X, then prove that the function $d(x, y) = \max\{d_1(x, y), d_2(x, y)\}$ defined on $X \times X$ is also a metric on X. What can you say about the minimum of two metrics?

1.3.3. Let (Y, ρ) be a non-empty metric space and X be a proper superset of Y (that is, $Y \subseteq X$). Extend the metric ρ to X, that is, define a metric d on X such that $d(y, y') = \rho(y, y')$ for all y, $y' \in Y$.

1.3.4. Use Example 1.3.6 to construct a metric on the extended real line $\mathbb{R} \cup \{+\infty, -\infty\}$.

1.3.5. Let $X = (0, \infty)$. Then verify that $d(x, y) = |\frac{1}{x} - \frac{1}{y}|$ for x, $y \in X$ is a metric on X.

1.3.6. Let $f : (X, d) \to (Y, \rho)$ be any function between two metric spaces. Show that $\sigma(x, x') = d(x, x') + \rho(f(x), f(x'))$ is also a metric on X.

1.3.7. Let (X, d) be a metric space. Prove that the following functions define a metric on X:

(i) $\rho(x, y) = \min\{1, d(x, y)\}$,

(ii) $\delta(x, y) = \frac{d(x,y)}{1+d(x,y)}$,

where x, $y \in X$. Do you observe some special property possessed by these metrics which need not be possessed by the metric d?

1.3.8. For $n \in \mathbb{N}$ and $f, g \in C(\mathbb{R})$, let $d_n(f,g) = \max_{x \in [-n,n]} |f(x) - g(x)|$. Further, let

$$d(f,g) = \sum_{n=1}^{\infty} \frac{1}{2^n} \frac{d_n(f,g)}{1 + d_n(f,g)}.$$

Show that d_n is a pseudometric on $C(\mathbb{R})$, while d is a metric on $C(\mathbb{R})$.

1.3.9. Let (X, d) be a metric space. Prove that the function ρ given by

$$\rho(x,y) = \ln(1 + d(x,y)),$$

is also a metric on X.

1.3.10. Consider the functions: $f(x) = x$ and $g(x) = 6 - x^2$ on $[0,3]$. Compute $D(f,g)$, $d_1(f,g)$ and $d_2(f,g)$. (Refer to Examples 1.3.11 and 1.3.12.)

1.4 Some Useful Concepts

Now we define the most important but a basic concept in a metric space (X, d). Consider first the metric space $(\mathbb{R}, |\cdot|)$. Given $x \in \mathbb{R}$ and $\delta > 0$, what does the open interval $(x - \delta, x + \delta)$ mean? Note

$$(x - \delta, x + \delta) = \{y \in \mathbb{R} : |x - y| < \delta\}.$$

So if we replace $\mathbb{R}$ by X and $|\cdot|$ by d, then the set $\{y \in \mathbb{R} : |x - y| < \delta\}$ is transformed into the set $\{y \in X : d(x,y) < \delta\}$ which is the collection of all points in X whose distance from x is less than δ. Now think about 3-dimensional Euclidean space $\mathbb{R}^3$ and the open ball centred at $x = (x_1, x_2, x_3)$ and radius δ. Keeping this example in mind, the set $\{y \in X : d(x,y) < \delta\}$ is denoted by $B(x, \delta)$ which is the open ball centred at x with radius δ.

Definitions 1.4.1. Let (X, d) be a metric space. For any $x \in X$ and any number $r > 0$, the set

$$B(x,r) = \{y \in X : d(x,y) < r\}$$

is called the **open ball** with centre x and radius r. Similarly, the set $\{y \in X : d(x,y) \leq r\}$ is called the **closed ball** with centre x and

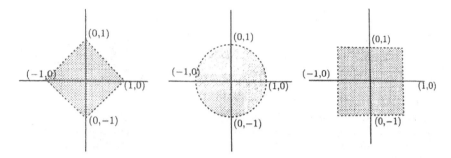

Figure 1.1 Open unit balls with respect to μ_1, μ_2 and μ_∞.

radius r and is denoted by $C(x, r)$. In particular, the balls $B(x, 1)$ and $C(x, 1)$ are said to be the *open unit ball* and *closed unit ball* respectively.

Evidently if $x \in (X, d)$ and $0 < s < r$, then $B(x, s) \subseteq B(x, r)$. But note that $B(x, s)$ may not be a proper subset of $B(x, r)$: in a metric space (X, d) with the discrete metric (Example 1.3.5), $B(x, 2) = B(x, 3) = X \ \ \forall \ x \in X$. *Can you draw geometrical figures for the open unit balls in $\mathbb{R}^2$ with respect to the metrics given in Example 1.3.4 (Figure 1.1)?*

Definitions 1.4.2. Let (X, d) be a metric space and let $A \subseteq X$. A point $a \in A$ is said to be an **interior point** of A if there exists $r > 0$ such that $B(a, r) \subseteq A$. The **interior** of A, denoted by int A or A^o, is the set of all interior points of A.

It should be noted that the radius of the ball $B(a, r)$ in Definition 1.4.2 may vary with the point $a \in A$.

Example 1.4.1. Consider the metric space $(\mathbb{R}, |\cdot|)$ and let $A = [0, 2)$. Note that every point in A except 0 is an interior point of A and hence int $A = (0, 2)$.

Example 1.4.2. Let (X, d) be a metric space with the discrete metric. If $x \in X$ and $r \leq 1$, then $B(x, r) = \{x\}$, while $B(x, r) = X$ if $r > 1$. Thus, int $A = A$ for $A \subseteq X$.

By the definition of an interior point, int $A \subseteq A$. In particular, int $\emptyset = \emptyset$, that is, the empty subset of X does not have any interior point. In Example 1.4.2, int $A = A$ for all $A \subseteq X$, while in Example 1.4.1, if we take $A = (0, 2)$, then int $A = A$ in $(\mathbb{R}, |\cdot|)$. In fact, the interior of an open interval is the interval itself. So you have two possibilities for a set A in a metric space (X, d) : (i) int $A = A$ or else (ii) int $A \subsetneq A$. The former case is an interesting one. Hence the subsets of a metric space possessing this property are known by a special name.

Definition 1.4.3. Let (X, d) be a metric space and A be a subset of X. Then A is called **open** in (X, d) if int $A = A$, that is, each point of A is an interior point of A.

Remark. Attention needs to be given on the metric space (X, d) in which the interior of a subset is considered. For example, the interior of $\mathbb{Q}$ in $(\mathbb{R}, |\cdot|)$ is the empty set $\emptyset$ (use the density of irrationals in $\mathbb{R}$), while the interior of $\mathbb{Q}$ in $(\mathbb{Q}, |\cdot|)$, since $(\mathbb{Q}, |\cdot|)$ itself is a metric space, is $\mathbb{Q}$. In fact, in general for any metric space (X, d), int X in (X, d) is X.

It is seen that the set of rational numbers $\mathbb{Q}$ can be considered in two ways: as a subset of $(\mathbb{R}, |\cdot|)$ or as a subset of $(\mathbb{Q}, |\cdot|)$. In a general scenario, let (X, d) be a metric space and Y be a non-empty subset of X. Then the metric d on X induces a metric d_Y on Y in the following natural manner: for y_1, $y_2 \in Y$, define $d_Y(y_1, y_2) = d(y_1, y_2)$. One can easily see that d_Y is a metric on Y. It is called the *metric on Y induced by the metric d* on X. Moreover, the metric space (Y, d_Y) is called a **metric subspace** of (X, d). The metric d_Y is essentially same as d, the only difference is that in case of d_Y the points are picked up from Y.

Note. *For convenience, the same notation d, in place of d_Y, is generally used to indicate the metric on Y induced by the metric d on X.*

If (X, d) is a metric space and Y is a non-empty subset of X, then one needs to be careful about the open balls in Y and that in X. Let $y \in Y$ and $r > 0$. Then the open ball $B(y, r)$ in Y is a subset of the open ball $B(y, r)$ in X. In order to maintain the difference, one should use the following different notations (if needed): $B_Y(y, r) = \{z \in Y : d(z, y) < r\}$ and $B_X(y, r) = \{z \in X : d(z, y) < r\}$. In other words, $B_Y(y, r) = B_X(y, r) \cap Y$. For example, consider $(\mathbb{R}, |\cdot|)$ and the subset $Y = [0, 2]$ of $\mathbb{R}$. Then $B_{\mathbb{R}}(0, \frac{1}{2}) = (-\frac{1}{2}, \frac{1}{2})$, while $B_Y(0, \frac{1}{2}) = [0, \frac{1}{2})$.

The basic facts about open sets are included in the following result.

Theorem 1.4.1. *Let (X, d) be a metric space. Then*

(a) X *and* $\emptyset$ *are open in* (X, d).
(b) *If* $\{U_i : i \in I\}$ *is an arbitrary family of open sets in* (X, d), *then their union* $\bigcup_{i \in I} U_i$ *is also open in* (X, d).
(c) *If* $\{U_1, U_2, \ldots, U_n\}$ *is a finite family of open sets in* (X, d), *then their intersection* $\bigcap_{i=1}^{n} U_i$ *is also open in* (X, d).

Proof. The proof of (a) is already discussed and the proof of (b) is left as an exercise. Here we prove (c):

Let $x \in \bigcap_{i=1}^{n} U_i$. Since $x \in U_i$ for $1 \leq i \leq n$ and int $U_i = U_i$, there exists $r_i > 0$ such that $x \in B(x, r_i) \subseteq U_i$. Let $r = \min_{1 \leq i \leq n} r_i$. Note that $r > 0$. Now $x \in B(x, r) \subseteq B(x, r_i) \subseteq U_i$ for all $i = 1, \ldots, n$ and hence $x \in B(x, r) \subseteq \bigcap_{i=1}^{n} U_i$. This implies that $\text{int}(\bigcap_{i=1}^{n} U_i) = \bigcap_{i=1}^{n} U_i$. Hence, $\bigcap_{i=1}^{n} U_i$ is open in (X, d). $\square$

The reader should observe that in Theorem 1.4.1(c), we have used the 'finiteness' in choosing r which is the minimum of all r_i's. Since only finitely many r_i's are there, $r > 0$. In case of an arbitrary family, we need to replace 'minimum' by 'infimum', that is, we need to choose $r = \inf_{i \in I} r_i$. But in this case, there is no guarantee that r will be strictly positive. For example, $\inf_{n \in \mathbb{N}} \frac{1}{n} = 0$, though $\frac{1}{n} > 0$ for all $n \in \mathbb{N}$. One may wonder that there might be other ways of proving Theorem 1.4.1(c) for an arbitrary family of open sets. But the following counter example shows that the result may not hold if we replace 'finite' by 'arbitrary'.

Example 1.4.3. Consider the infinite family of open sets $\{(-\frac{1}{n}, \frac{1}{n}) : n \in \mathbb{N}\}$ in $\mathbb{R}$. Then $\bigcap_{n \in \mathbb{N}} (-\frac{1}{n}, \frac{1}{n}) = \{0\}$ which is not open in $\mathbb{R}$.

Analogous to interior points and open sets, one can also define closure points of a set A in a metric space and then with the help of closure points, closed sets in a metric space (X, d) can be defined. Or else, we can define a closed set first and then we link it to the closure points. Let us take the latter route.

Definition 1.4.4. Let (X, d) be a metric space. A subset C of X is called **closed** in (X, d) if $X - C$ is open in (X, d), where $X - C$ is the complement of C in X.

Remarks. (a) A door of a room is either open or closed. But a set
A in a metric space may be neither open nor closed. Look at the
set $[0, 2)$ in $(\mathbb{R}, |\cdot|)$ — it is neither open nor closed in $(\mathbb{R}, |\cdot|)$.
(b) Since X and $\emptyset$ are open in a metric space (X, d), they are also
closed in (X, d). That is, X and $\emptyset$ are both open and closed in a
metric space (X, d).

When a set is both open and closed in a metric space (X, d), it is
called **clopen** in a metric space (X, d). So, X and $\emptyset$ are always clopen
in a metric space (X, d). Now the question arises, whether in general
there exists a third set, other than X and $\emptyset$, which is both open
and closed? This question needs to be answered as it is related to
connectedness of a metric space which will be studied in Chapter 7.
So consider a metric space (X, d) where d is the discrete metric. It
is known that every subset of X is open in (X, d) and hence every
subset is also closed in (X, d). So if X has more than one point, then
in the metric space (X, d), we can have a third clopen set as well.

In a metric space (X, d), a subset A of X is open in (X, d) if
and only if $X - A = A^c$ is closed in (X, d). So the concepts of open
and closed sets in a metric space are dual. Now the following result
follows immediately from Theorem 1.4.1 and De Morgan's laws.

Theorem 1.4.2. *For a metric space (X, d) the following statements
hold:*

(i) *X and $\emptyset$ are closed in (X, d).*
(ii) *Arbitrary intersections of closed sets in (X, d) is again closed in
(X, d).*
(iii) *Finite unions of closed sets in (X, d) is again closed in (X, d).*

Now we define the closure points of a set A in a metric space
(X, d). Think about the word 'closure' (it is an adjective to a (noun)
point). Intuitively, a point x in X should be called a closure point of A
if x is really close to A. Note that if x is already in A, then obviously
x is close to A. But what about the points which are not in A, but
'really' close to A. Let us look at the following example: consider the
set $A = (0, 2)$ in the metric space $(\mathbb{R}, |\cdot|)$. Consider two points 0
and 2. They are not in A. But if you take any ϵ-neighbourhood of 0,
$(-\epsilon, \epsilon)$, it must intersect A, no matter how small ϵ you choose. On
the other hand, -0.01 does not belong to A but a sufficiently small
$\epsilon > 0$ can be chosen such that $(-0.01 - \epsilon, -0.01 + \epsilon) \cap A = \emptyset$ (take

$\epsilon < 0.01$). So clearly with respect to the set A, there is a difference between the points 0 and -0.01. A similar situation occurs with the point 2 as well. So though 0 and 2 do not belong to A, but they possess some special property which is not possessed by other points in A^c. One can say that 0 and 2 are closure points of A. Now we give a precise definition of a closure point. But first let us define neighbourhood ('nhood' in short) formally.

Definition 1.4.5. Let (X, d) be a metric space and $x \in X$. Then a **neighbourhood** of x in (X, d) is an open set in (X, d) that contains x.

If U is a nhood of x in (X, d), then by definition, it contains an open ball around x, that is, there exists some $r > 0$ such that $B(x, r) \subseteq U$. Consequently, while dealing with nhoods, often it is enough to work with the open balls. Hence they are given a special name. Any open ball with centre x in (X, d) is called a **basic neighbourhood** of x in (X, d).

Definitions 1.4.6. Let (X, d) be a metric space, $A \subseteq X$ and $x \in X$. Then x is called a **closure point** of A if $B(x, r) \cap A \neq \emptyset$ for all $r > 0$, that is, every nhood of x intersects A. The set of all closure points of A in (X, d) is called the **closure** of A and is denoted by $\mathrm{cl}_X A$ or $\overline{A}$.

Note that by definition $A \subseteq \overline{A}$. So a natural question is: when do we have the reverse inclusion? The next result answers more than this.

Theorem 1.4.3. *Let (X, d) be a metric space and $A \subseteq X$. Then $\overline{A}$ is a closed set. Moreover, $\overline{A}$ is the smallest closed set in (X, d) containing A, that is, if $A \subseteq F$ and F is closed in (X, d), then $\overline{A} \subseteq F$.*

Proof. First we show that $\overline{A}$ is closed, that is, $X - \overline{A}$ is open in (X, d). So we need to show that every point of $X - \overline{A}$ is an interior point of $X - \overline{A}$. So let $x \in X - \overline{A}$. Then there exists $r > 0$ such that $B(x, r) \cap A = \emptyset$. This means $B(x, r) \subseteq X - A$. But we will show that $B(x, r) \subseteq X - \overline{A}$, that is, $B(x, r) \cap \overline{A} = \emptyset$. Let $y \in B(x, r)$. Then $r - d(x, y) > 0$. Choose $0 < s < r - d(x, y)$. Then using triangle's inequality, one can prove that $B(y, s) \subseteq B(x, r)$. Since $B(x, r) \cap A = \emptyset$, $B(y, s) \cap A = \emptyset$. This implies that y is not a closure point of A, that is, $y \notin \overline{A}$, $y \in X - \overline{A}$. But y was arbitrarily chosen from $B(x,r)$. Hence $B(x,r) \subseteq X - \overline{A}$. Thus, x is an interior point of $X - \overline{A}$. Consequently, $X - \overline{A}$ is open, which implies that $\overline{A} = X - (X - \overline{A})$ is closed in (X,d).

For the second part, let F be a closed subset of (X,d) containing A. Thus $X - F$ is open in (X,d), which implies that every point of $X - F$ is an interior point of $X - F$. So given $x \in X - F$, there exists $r > 0$ such that $B(x,r) \subseteq X - F$, that is, $B(x,r) \cap F = \emptyset$. Since $A \subseteq F$, $B(x,r) \cap A = \emptyset$. But this means that x is not a closure point of A, that is, $x \in X - \overline{A}$. Hence $X - F \subseteq X - \overline{A}$, that is, $\overline{A} \subseteq F$. $\square$

Remark. Since an arbitrary intersection of closed sets is closed, $C = \bigcap \{F \subseteq X : F$ is closed in (X,d) and $A \subseteq F\}$ is the smallest closed set containing A. *Just by using the definition of a closure point, can you prove that $C = \overline{A}$?*

Similarly, we can have the following result.

Theorem 1.4.4. *Let (X,d) be a metric space and $A \subseteq X$. Then int A is the largest open set contained in A.*

In fact, interior and closure are dual concepts. The precise statement is given in the next lemma.

Lemma 1.4.1. *Let (X,d) be a metric space and $A \subseteq X$. Then int $A = X - (\overline{X - A})$, that is, $X-$ int $A = \overline{X - A}$.*

Now we will discuss a special kind of closure point known as accumulation point. Let us start with an example. Consider the metric space $(\mathbb{R}, |\cdot|)$ and $A = (0,1) \cup \{2\}$. Note that $\overline{A} = [0,1] \cup \{2\}$. Though 0 and 1 do not belong to A, there is a remarkable difference between these two points $\{0,1\}$ on one side and the point 2 on the other side. Pick up the pair $\{0,2\}$. Discussion for the pair $\{1,2\}$ is similar. Both 0 and 2 are closure points of A. But if you take any ϵ-nhood $(-\epsilon, \epsilon)$ of 0 in $\mathbb{R}$, it will intersect A in infinitely many points. In particular, $((-\epsilon, \epsilon) - \{0\}) \cap A \neq \emptyset$. But this is not true for 2. If we choose $0 < \epsilon < 1$, then $(2 - \epsilon, 2 + \epsilon) \cap A = \{2\}$, that is, $(2 - \epsilon, 2 + \epsilon)$ does not intersect A in points other than 2, that is, $((2 - \epsilon, 2 + \epsilon) - \{2\}) \cap A = \emptyset$. Also note that $((-\epsilon, \epsilon) - \{0\}) \cap A = (-\epsilon, \epsilon) \cap (A - \{0\})$. In general, for a metric space (X,d), $x \in X$ and $A \subseteq X$, $(B(x,r) - \{x\}) \cap A = B(x,r) \cap (A - \{x\})$ for all $r > 0$. In view of this discussion, we make the following definition of accumulation points.

Definition 1.4.7. Let (X, d) be a metric space, $A \subseteq X$ and $x \in X$. Then x is called an **accumulation point** of A in (X, d) if $B(x, r) \cap (A - \{x\}) \neq \emptyset$ for all $r > 0$, that is, every nhood of x contains an element of A distinct from x. Moreover, the set of all accumulation points of A in (X, d) is called the **derived set** of A and is denoted by A'.

Remark. Note that some authors use the term 'limit point' in place of accumulation point.

Definition 1.4.8. Let (X, d) be a metric space, $A \subseteq X$ and $x \in A$. Then x is called an **isolated point** of A in (X, d) if there exists $r_o > 0$ such that $B(x, r_o) \cap (A - \{x\}) = \emptyset$, that is, there exists a nhood of x which does not contain any element of A other than x itself.

In the example discussed before these definitions, every point in $[0, 1]$ is an accumulation point of $A = (0, 1) \cup \{2\}$ in $(\mathbb{R}, |\cdot|)$. Note that though 0 and 1 are accumulation points of A, they do not belong to A. On the other hand, although 2 is in A, it is not an accumulation point of A but it is an isolated point of A. Evidently, if x is an element of $A \subseteq X$, then x is an isolated point of A if and only if x is not an accumulation point of A.

The proof of the next result is easy and hence left as an exercise for the readers.

Theorem 1.4.5. *Let (X, d) be a metric space and $A \subseteq X$. Then*

(a) A *is closed in* (X, d) $\Leftrightarrow$ $A = \overline{A}$.
(b) $\overline{A} = A \cup A'$

It is evident that the closure of a set A in a metric space (X, d) is a subset of X. But when $\overline{A}$ coincides with the mother space X, then A plays a special role in X.

Definition 1.4.9. Let (X, d) be a metric space and $\emptyset \neq A \subseteq B \subseteq X$. Then A is said to be **dense** in B if every point of B is a closure point of A in X, that is, if $B \subseteq \overline{A}$, where $\overline{A}$ denotes the closure of A in X. In particular, A is dense in X if $\overline{A} = X$.

Example 1.4.4. Both $\mathbb{P}$ = the set of irrationals and $\mathbb{Q}$ = the set of rationals are dense in $(\mathbb{R}, |\cdot|)$. Similarly, one can see that the set $\{(a,b) \in \mathbb{R}^2 : a, \ b \in \mathbb{Q}\}$ is dense in $\mathbb{R}^2$.

In addition to being dense in $(\mathbb{R}, |\cdot|)$, the set of rationals $\mathbb{Q}$ is also countable which makes it a special subset of $(\mathbb{R}, |\cdot|)$. With this motivation, let us look for such subsets in general metric spaces as well. In this regard, another class of metric spaces is now defined.

Definition 1.4.10. Let (X, d) be a metric space and $\emptyset \neq B \subseteq X$. If there exists a countable dense subset of B, then B is called **separable** (in (X, d)).

Example 1.4.5. The metric space $(\mathbb{R}, |\cdot|)$ is separable. Similarly, one can easily prove that the complex plane $\mathbb{C}$ with the usual metric is separable: the set $A = \{a + \iota b : a, \ b \in \mathbb{Q}\}$ is a countable dense subset of $\mathbb{C}$.

Example 1.4.6. The l^p-space with $1 \leq p < \infty$ is separable. Consider the set,

$$B = \{(r_1, \ r_2, \ldots, r_n, 0, 0, \ldots) : n \in \mathbb{N}, \ r_i \in \mathbb{Q} \text{ for } 1 \leq i \leq n\}.$$

Then B is a countable subset of l^p because finite product of countable sets is countable and countable union of countable sets is also countable. Now we show that B is dense in l^p. Let $x = (\eta_j)$ be an arbitrary element in l^p and $\epsilon > 0$. Then there exists $n_o \in \mathbb{N}$ such that

$$\sum_{j=n_o+1}^{\infty} |\eta_j|^p < \frac{\epsilon^p}{2}.$$

Since the set of rationals is dense in $\mathbb{R}$, for each η_j where $1 \leq j \leq n_o$, there exists $r_j \in \mathbb{Q}$ such that

$$\sum_{j=1}^{n_o} |\eta_j - r_j|^p < \frac{\epsilon^p}{2}.$$

This implies that

$$(d(x,y))^p = \sum_{j=1}^{n_o} |\eta_j - r_j|^p + \sum_{j=n_o+1}^{\infty} |\eta_j|^p < \epsilon^p,$$

where $y = (r_1, \ r_2, \ldots, r_{n_o}, 0, 0, \ldots) \in B$. Consequently, $d(x,y) < \epsilon$ and hence B is dense in l^p.

Example 1.4.7. The l^∞-space is not separable: the set of all sequences of zeros and ones is an uncountable set by Cantor's diagonal method. Any two members of this set are distance 1 apart in l^∞. Consequently, l^∞ cannot contain a countable dense subset. *Now can you similarly prove that the space $B(X)$ is not separable for any infinite set X?*

The routine proof of the next result is omitted.

Proposition 1.4.1. *Let (X,d) be a metric space and $\emptyset \neq A \subseteq B \subseteq X$. Then A is dense in B if and only if A is dense in $\overline{B}$, where $\overline{B}$ denotes the closure of B in (X,d). In particular, B is separable if and only if $\overline{B}$ is separable.*

Definition 1.4.11. Let (X,d) be a metric space and $A \subseteq X$. Then the boundary of A is defined to be the set $\overline{A} \cap \overline{X-A}$ and it is denoted by ∂A. A point in ∂A is called a **boundary point** of A.

Note that a point x in X is a boundary point of A if and only if every nhood of x intersects both A and $X-A$. Now consider $\mathbb{Q} \subseteq \mathbb{R}$, then $\partial \mathbb{Q} = \mathbb{R}$ but if $\mathbb{Q}$ is considered as a subset of itself then it has no boundary point. In general, $\partial X = \emptyset$ for any metric space (X,d).

Remarks. (a) $\partial A = \partial(X-A)$.
(b) Often in literature, in place of 'boundary', the term 'frontier' is also used.
(c) ∂A is always a closed set.
(d) $\partial A = \overline{A} \setminus A^\circ$.

Now consider the interval $[0,1)$ in $\mathbb{R}$. Observe that if $[0,1)$ is considered as a subset of $[0,1]$, then only 1 is the boundary point of $[0,1)$. But if $[0,1)$ is considered as a subset of $\mathbb{R}$ then 0 and 1 are the boundary points of $[0,1)$. Thus while talking about the boundary points of a set A, one needs to be careful about the metric space of which A is considered as a subset.

When (X,d) is a metric space and $\emptyset \neq Y \subseteq X$, an open set A in (Y,d) need not be open in (X,d). The open ball in (Y,d) centred at y with radius r, $B_Y(y,r) = B_X(y,r) \cap Y$ may not be open in (X,d). For example, if we consider the metric space $(\mathbb{R}, |\cdot|)$ and $Y = [0,1]$, then $[0,\frac{1}{2})$ is open in $(Y, |\cdot|)$, but not open in $(\mathbb{R}, |\cdot|)$. When we want to emphasize the open and closed sets in the metric subspace

(Y, d), we use the terms relatively open and relatively closed in Y. The precise definition follows.

Definition 1.4.12. Let (X, d) be a metric space and $\emptyset \neq Y \subseteq X$. If $U \subseteq Y$ and U is open in the metric subspace (Y, d), then U is called **relatively open** in Y. Similarly, if U is closed in (Y, d), then U is called **relatively closed** in Y.

Proposition 1.4.2. *Let (X, d) be a metric space and $\emptyset \neq Y \subseteq X$.*

(a) *A subset U of Y is relatively open in Y if and only if there exists an open set V in (X, d) such that $U = V \cap Y$.*
(b) *A subset C of Y is relatively closed in Y if and only if there exists a closed set D in (X, d) such that $C = D \cap Y$.*

Proof. (a) Let U be open in (Y, d). Then for each $y \in U$, there exists $r_y > 0$ such that $y \in B_Y(y, r_y) \subseteq U$. Hence $U = \bigcup_{y \in U} B_Y(y, r_y)$. Let $V = \bigcup_{y \in U} B_X(y, r_y)$. Then V is open in (X, d) and

$$V \cap Y = \left(\bigcup_{y \in U} B_X(y, r_y) \right) \cap Y = \bigcup_{y \in U} B_X(y, r_y) \cap Y$$

$$= \bigcup_{y \in U} B_Y(y, r_y) = U.$$

Similarly, we can prove the converse.
(b) If C is closed in (Y, d), then by (a), there exists an open set V in (X, d) such that $Y - C = V \cap Y$. Then $C = (X - V) \cap Y$. But $X - V$ is closed in (X, d). Take $D = X - V$. Similarly, we can prove the converse part. $\square$

It is straightforward to verify that if $A \subseteq Y \subseteq X$, then $\mathrm{cl}_Y(A) = Y \cap \mathrm{cl}_X(A)$, where $\mathrm{cl}_Y(A)$ denotes the closure of A in (Y, d) and $\mathrm{cl}_X(A)$ denotes the closure of A in (X, d).

In the upcoming chapters, we will see that certain properties of metric spaces can be completely defined in terms of open sets (for example, continuity and compactness). Consequently, the rich structure of a 'metric' is not required in those cases. Hence a special name is given to the collection of all open subsets of a metric space. It is called the *topology on X induced by the metric d*. If you carefully observe, we mainly require the properties (of open sets)

listed in Theorem 1.4.1 for the study. This gives an idea for defining
the notion of topology on an arbitrary set.

Definitions 1.4.13. Let X be a set and τ be a family of subsets of
X satisfying the following properties:

(a) $\emptyset \in \tau$ and $X \in \tau$.
(b) if $\{A_i : i \in I\} \subseteq \tau$, then $\bigcup_{i \in I} A_i \in \tau$.
(c) if $A_1, \ldots, A_n \in \tau$, then $\bigcap_{k=1}^{n} A_k \in \tau$.

Then the family τ^1 is called a **topology** on X and the pair (X, τ) is
called a **topological space**. Moreover, the elements of τ are called
open sets.

   Note that every topology need not be induced by a metric. Since
the study of topological spaces is like exploring another world, we
won't be going into further details of topological spaces. But of
course, we would like to advise our readers that after studying metric
spaces they should also explore the bigger world of topology.

Exercises

1.4.1. Compute the interior, closure, boundary and the derived set
of

(a) $A = \mathbb{Q}^c \cap [0, \sqrt{3})$ in $(\mathbb{Q}^c, |\cdot|)$.
(b) $B = \{(x, y) \in \mathbb{R}^2 : x + y < 1, \ x \geq 0, \ y \geq 0\}$ in $(X, |\cdot|)$ where
    $X = \{(x, y) \in \mathbb{R}^2 : x \geq 0, \ y \geq 0\}$.

1.4.2. Let (X, d) be a metric space, $x \in X$ and $r > 0$.

(a) Show that $B(x, r)$ is an open set in $(X \cdot d)$.
(b) Show that $C(x, r)$ is a closed set in $(X \cdot d)$.

1.4.3. Let A and B be subsets of a metric space. Show that

(a) $(A \cap B)^o = A^o \cap B^o$.
(b) $A^o \cup B^o \subseteq (A \cup B)^o$.
(c) $\overline{A \cup B} = \overline{A} \cup \overline{B}$.
(d) $\overline{A \cap B} \subseteq \overline{A} \cap \overline{B}$.

---

[1]A Greek letter pronounced as tau.

(e) $\overline{A} = (\text{int}(A^c))^c$ and $A^\circ = (\text{cl}(A^c))^c$.

Parts (b) and (d) are also true in case we take an arbitrary collection of subsets but the equality in (a) and (c) may not hold for an arbitrary collection. Think of counter examples.

1.4.4. If A and B are subsets of a metric space such that $A \subseteq B$, then show that $\overline{A} \subseteq \overline{B}$. Is the converse also true, that is, if $\overline{A} \subseteq \overline{B}$, does this imply that $A \subseteq B$?

1.4.5. Suppose $\emptyset \neq A \subseteq B \subseteq X$, where (X, d) is a metric space. Then show that A is dense in B if and only if $\text{cl}_B A = B$, where $\text{cl}_B A$ denotes the closure of A in (B, d).

1.4.6. Let (X, d) be a metric space and x and y be two distinct points in X. Show that there exists $r > 0$ such that $B(x, r) \cap B(y, r) = \emptyset$.

1.4.7. Let (X, d) be a metric space and U be a non-empty subset of X. Show that U is open in (X, d) if and only if U can be written as a union of a family of open balls. In view of this problem, an open set need not be an open ball — give an example. Similarly, a closed set in (X, d) need not be a closed ball.

1.4.8. Let (X, d_X) and (Y, d_Y) be metric spaces and equip $X \times Y$ with the metric μ_∞ (refer to Example 1.3.4). Show that if A and B are open (resp. closed) subsets of X and Y, then $A \times B$ is open (resp. closed) in $(X \times Y, \mu_\infty)$. Also see Exercise 2.6.7.

1.4.9. Let (X, d) be a metric space and A and B be two subsets of X. If A is open and B is closed in (X, d), then show that $A - B$ is open in (X, d), while $B - A$ is closed in (X, d).

1.4.10. If A is a subset of a metric space (X, d), then show that $\partial A = \emptyset$ if and only if A is both open and closed in X.

1.4.11. Consider $A = \{(x, \sin(\frac{1}{x})) : x > 0\}$ as a subset of $\mathbb{R}^2$. Then show that $A \cup \{(0, y) : y \in [-1, 1]\}$ is the boundary of A in $\mathbb{R}^2$.

1.4.12. Show that a metric space (X, d) with discrete metric is separable if and only if X is countable.

1.4.13. What can you say regarding the separability of $C_b(\mathbb{R})$, where $C_b(\mathbb{R})$ is the metric subspace of $B(\mathbb{R})$, which consists of all the continuous and bounded functions defined on $\mathbb{R}$?

1.4.14. Show that any non-empty subset of a separable metric space is separable.

1.4.15. Let D_i be a dense subset of the metric space (X_i, d_i) for all $i \in \{1, 2, \ldots, n\}$. Show that $\prod_{i=1}^{n} D_i$ is dense in $(\prod_{i=1}^{n} X_i, d)$, where d is the metric defined as: $d(x, y) = \sum_{i=1}^{n} d_i(x_i, y_i)$ for $x = (x_1, \ldots, x_n)$ and $y = (y_1, \ldots, y_n)$ in $\prod_{i=1}^{n} X_i$. Hence deduce that the product of a finite number of separable metric spaces is also separable.

1.4.16. Let $\{(X_i, d_i) : i \in \mathbb{N}\}$ be a countable collection of metric spaces. Then show that the Cartesian product $\prod_{i=1}^{\infty} X_i$ equipped with the metric given in Example 1.3.10 is separable if and only if (X_i, d_i) is separable for all $i \in \mathbb{N}$.

1.5 Convergence of Sequences

Recall that in order to define a metric d on a non-empty set, we took the help of $(\mathbb{R}, |\cdot|)$. A sequence (x_n) is said to converge to x in $\mathbb{R}$ if given $\epsilon > 0$, there exists $n_\epsilon \in \mathbb{N}$ (n_ϵ depends on ϵ) such that $|x_n - x| < \epsilon$ for all $n \geq n_\epsilon$, that is, $x_n \in (x - \epsilon, x + \epsilon)$ for all $n \geq n_\epsilon$. Note that $(x - \epsilon, x + \epsilon)$ is a nhood of x. But $\epsilon > 0$ was chosen arbitrarily. So if we imitate this definition for a general metric space (X, d), we get the following definition. But before that, note that a sequence (x_n) in X is just a function $f : \mathbb{N} \to X$ and $f(n)$ is denoted by x_n.

Definition 1.5.1. Let (X, d) be a metric space, $x \in X$ and (x_n) be a sequence in X. Then (x_n) is said to **converge** to x in (X, d) if for every $\epsilon > 0$, there exists $n_\epsilon \in \mathbb{N}$ (n_ϵ depends on ϵ) such that $d(x_n, x) < \epsilon$ for all $n \geq n_\epsilon$, that is, $x_n \in B(x, \epsilon)$ for all $n \geq n_\epsilon$. The point x is called a **limit** of the sequence (x_n) in (X, d).

Evidently, if (x_n) converges to x in (X, d) then the corresponding real sequence $\{d(x_n, x)\}$ converges to 0 in $\mathbb{R}$.

If x is a limit of the sequence (x_n), then it is often denoted as: $x_n \to x$ or $\lim_{n \to \infty} x_n = x$. In the previous definition, we can call a limit of a convergent sequence to be *the* limit (as proved in the next result).

Proposition 1.5.1. *Let (X,d) be a metric space and (x_n) be a sequence in X. If (x_n) converges to x and y in X, then $x = y$.*

Proof. Let (x_n) be a sequence in (X,d) which converges to x and y in X. Then for $\epsilon > 0$, there exists $n_o \in \mathbb{N}$ such that for $n \geq n_o$, we have

$$d(x,y) \leq d(x,x_n) + d(x_n,y) < \epsilon.$$

Since ϵ was arbitrary and $d(x,y) \geq 0$, $x = y$. $\qquad\square$

Sequences play a crucial role in the theory of metric spaces because many of the concepts can be conveniently handled with the help of sequences. Let us start with the following sequential characterization of closure point and accumulation point in (X,d).

Theorem 1.5.1. *Let (X,d) be a metric space, $x \in X$ and A be a non-empty subset of X. Then*

(a) $x \in \overline{A}$ *(that is, x is a closure point of A) if and only if there exists a sequence (x_n) in A such that $x_n \to x$ in (X,d).*

(b) $x \in A'$ *(that is, x is an accumulation point of A) if and only if there exists a sequence (x_n) of distinct terms in A such that $x_n \to x$ in (X,d).*

Proof. (a) First suppose that $x \in \overline{A}$. Hence by definition $B(x,\frac{1}{n}) \cap A \neq \emptyset$. Choose $x_n \in B(x,\frac{1}{n}) \cap A$. Then clearly (x_n) is in A and $d(x,x_n) < \frac{1}{n}$ which implies that $x_n \to x$ in (X,d).

Conversely, suppose that there exists a sequence (x_n) in A such that $x_n \to x$ in (X,d). So given $\epsilon > 0$, there exists $n_\epsilon \in \mathbb{N}$ such that $x_n \in B(x,\epsilon)$ for all $n \geq n_\epsilon$. In particular, $B(x,\epsilon) \cap A \neq \emptyset$. Since $\epsilon > 0$ was chosen arbitrarily, $x \in \overline{A}$.

(b) Let $x \in A'$. In the proof of the corresponding part in (a), we just chose x_n from $B(x,\frac{1}{n})$. But there was no guarantee that this x_n was different from x_m chosen from $B(x,\frac{1}{m})$. This time, we need to ensure it.

For $n = 1$, choose $x_1 \in B(x,1) \cap A$ such that $x \neq x_1$. This is possible because x is an accumulation point of A. For the next step we need to choose x_2 from $B(x,\frac{1}{2}) \cap A$ such that $x_2 \neq x_1$

and $x_2 \neq x$. Since $x_1 \neq x$, $d(x_1, x) > 0$. Choose r_2 such that $0 < r_2 < \min\{\frac{1}{2}, d(x_1, x)\}$. Then since x is an accumulation point of A, $B(x, r_2) \cap (A \setminus \{x\}) \neq \emptyset$. Choose $x_2 \in B(x, r_2) \cap (A \setminus \{x\})$.

Clearly, $x_2 \neq x$. Also $d(x_2, x) < r_2 < d(x_1, x)$ which implies $x_2 \neq x_1$. Now apply an inductive process. Suppose $x_1, x_2, \ldots, x_n$ are already chosen from A which are all distinct and $x \neq x_i$ for $1 \leq i \leq n$ such that $d(x, x_i) < \min\{\frac{1}{i}, d(x_{i-1}, x)\}$ for $2 \leq i \leq n$. Then choose $r_{n+1} > 0$ such that $r_{n+1} < \min\{\frac{1}{n+1}, d(x_n, x)\}$.

Then $B(x, r_{n+1}) \cap (A \setminus \{x\}) \neq \emptyset$ and choose $x_{n+1} \in B(x, r_{n+1}) \cap (A \setminus \{x\})$. Thus, (x_n) is a sequence in A with distinct terms which converges to x in (X, d).

The converse part clearly follows from the definition. □

Corollary 1.5.1. *If A is a subset of a metric space (X, d), then A is closed in X if and only if whenever $(a_n) \subseteq A$ converges to $x \in X$ then $x \in A$.*

As a consequence of the previous corollary, one can easily deduce that every finite subset of a metric space (X, d) is closed in X.

Corollary 1.5.2. *Let A be a subset of a metric space (X, d), then the derived set A' is a closed set.*

Corollary 1.5.3. *Let (X, d) be a metric space, $A \subseteq X$ and $x \in X$. Then x is an accumulation point of A in (X, d) if and only if for every open set U containing x, $U \cap A$ is infinite.*

Before we mention the next corollary, recall that $(x_{k_n})_{n \in \mathbb{N}}$ is a subsequence of $(x_n)_{n \in \mathbb{N}}$ in (X, d) if (k_n) is a strictly increasing sequence of natural numbers, that is, $k_n \in \mathbb{N}$ and $k_n < k_{n+1}$ for all $n \in \mathbb{N}$. More precisely, if the function $f : \mathbb{N} \to X$ defined by $f(n) = x_n$ represents the sequence (x_n), then a subsequence (x_{k_n}) of (x_n) is the composition function $f \circ \phi : \mathbb{N} \to X$ where $\phi : \mathbb{N} \to \mathbb{N} : \phi(n) = k_n$ is a strictly increasing function (that is, $\phi(n) < \phi(n+1) \ \forall \ n \in \mathbb{N}$) and $(f \circ \phi)(n) = f(\phi(n)) = x_{\phi(n)} = x_{k_n}$. Thus, (x_{k_n}) is itself a sequence.

Corollary 1.5.4. *Let (X, d) be a metric space and (x_n) be a sequence of distinct points in X. If $A = \{x_n : n \in \mathbb{N}\}$ has an accumulation point x in X, then there exists a subsequence (x_{k_n}) of (x_n) such that $x_{k_n} \to x$ in (X, d).*

Proof. By Corollary 1.5.3, $B(x,1) \cap A$ is infinite. So we can pick up $x_{k_1} \in B(x,1) \cap A$ such that $x_{k_1} \neq x$. Now $B(x,\frac{1}{2}) \setminus \{x_{k_1}\}$ is an open set containing x and hence $(B(x,\frac{1}{2}) \setminus \{x_{k_1}\}) \cap A$ is again infinite. So we can choose $x_{k_2} \in (B(x,\frac{1}{2}) \setminus \{x_{k_1}\}) \cap A$ such that $k_2 > k_1$ and $x_{k_2} \neq x$. Note that $d(x_{k_2}, x) < \frac{1}{2}$. Now continue this process inductively. Suppose that $x_{k_1}, \ldots, x_{k_n}$ have been chosen from A such that $k_1 < k_2 < \cdots < k_n$, $d(x_{k_j}, x) < \frac{1}{j}$ and $x_{k_j} \neq x$ for $j = 1, 2, \ldots, n$. Then consider the open set $B(x, \frac{1}{n+1}) \setminus \{x_{k_1}, \ldots, x_{k_n}\}$ containing x. Thus $(B(x, \frac{1}{n+1}) \setminus \{x_{k_1}, \ldots, x_{k_n}\}) \cap A$ is infinite. So we can choose $x_{k_{n+1}} \in B(x, \frac{1}{n+1}) \setminus \{x_{k_1}, \ldots, x_{k_n}\}$ such that $k_{n+1} > k_n$ and $x_{k_{n+1}} \neq x$. Now clearly, (x_{k_n}) is a subsequence of (x_n) such that $x_{k_n} \to x$ in (X,d). $\square$

Corollary 1.5.5. *Let (X,d) be a metric space and A be an infinite subset of X. Suppose every infinite subset of A has an accumulation point in A. Then every sequence (x_n) in A has a subsequence (x_{k_n}) converging to a point in A.*

Proof. Let (x_n) be a sequence in A and $B = \{x_n : n \in \mathbb{N}\}$. If B is finite, then (x_n) has a constant subsequence and clearly this constant subsequence converges to a point in A. So assume that $B = \{x_n : n \in \mathbb{N}\}$ is infinite. Then inductively we can prove the existence of a subsequence (x_{k_n}) of (x_n) such that $x_{k_n} \neq x_{k_m}$ holds whenever $n \neq m$. Let $A_1 = \{x_{k_n} : n \in \mathbb{N}\}$. Since A_1 is infinite, by hypothesis, A_1 has an accumulation point x (say) in A. Then by Corollary 1.5.4, there exists a subsequence $(x_{k_{n_j}})$ of (x_{k_n}) such that $x_{k_{n_j}} \to x$. Since (x_{k_n}) is a subsequence of (x_n), $(x_{k_{n_j}})$ is also a subsequence of (x_n). $\square$

Convergence of a sequence is a very strong property which is not possessed by every sequence. So there was a need to look for some weaker class of sequences which could be helpful in some sense. If we observe any convergent sequence carefully, then we will find that the sequential terms eventually comes closer to each other. This property is formalized under the name 'Cauchy sequence'.

Definition 1.5.2. Let (X,d) be a metric space and (x_n) be a sequence in X. Then (x_n) is called a **Cauchy sequence** in (X,d)

if for every $\epsilon > 0$, there exists $n_\epsilon \in \mathbb{N}$ (n_ϵ depends on ϵ) such that $d(x_m, x_n) < \epsilon$ for all $n, m \geq n_\epsilon$.

It is easy to see that every convergent sequence is Cauchy. But the converse is not in general true. The metric spaces in which the converse is also true holds a special place in the analysis on metric spaces.

Definition 1.5.3. If every Cauchy sequence in (X, d) converges in (X, d), then (X, d) is called a **complete metric space**.

Interestingly, $(\mathbb{R}, | \cdot |)$ is a complete metric space but $(\mathbb{Q}, | \cdot |)$ is not complete. We will have a detailed discussion on complete metric spaces in Chapter 3.

So far in a metric space, we were discussing about distance between two points, but at times we are prompted to expand this vision further and linking this with distance between a point and a set. Let (X, d) be a metric space, $\emptyset \neq A \subseteq X$ and $x \in X$. *How to define the distance between the point x and the set A?* First consider the Euclidean space $\mathbb{R}^3$. Suppose you are in front of the door of a room. How will you compute the distance between you and the room? Again in $\mathbb{R}^2$ with the usual distance, take a straight line L and a point (x, y) outside L. What is the distance between (x, y) and L? Will you go for the orthogonal projection of the point (x, y) on L? Suppose this projection meets L in the point (x', y'). So we take the distance to be the length of the line segment joining (x, y) and (x', y'). But the concept of orthogonal projection may not be applicable in a general metric space. Thus we look for some characteristic feature of the orthogonal projection which could be generalized as well. If we join (x, y) to each point on L and consider the lengths of the line segments, then the length of the line segment corresponding to the orthogonal projection is the smallest among all these lengths. In view of this observation, we have the following definition.

Definitions 1.5.4. Let (X, d) be a metric space, A and B be two non-empty subsets of X and $x \in X$. Then the distance of x from A, denoted as $d(x, A)$, is defined by

$$d(x, A) = \inf\{d(x, a) : a \in A\}.$$

Also, the gap between the sets A and B is given by

$$d(A, B) = \inf\{d(a, b) : a \in A, \ b \in B\}.$$

Note. *By convention,* $d(x, \emptyset) = +\infty$.

Suppose $q \in \mathbb{Q}$, then due to density of $\mathbb{Q}$ in $\mathbb{R}$, $d(q, \mathbb{Q} \setminus \{q\}) = 0$. Furthermore, let $A = \{(x, \frac{1}{x}) : x > 0\}$ and $B = \{(x, 0) : x \in \mathbb{R}\}$. Then note that A and B are disjoint closed subsets in $\mathbb{R}^2$, but $d(A, B) = 0$. Consequently, the function d does not define a metric on the collection of all non-empty subsets of a metric space (X, d). So the next guess is: take *supremum* instead of *infimum*. Moreover, since we need a metric to be real-valued, we should restrict ourselves to the collection of all non-empty bounded subsets of (X, d). Let $f(A, B) = \sup\{d(a, B) : a \in A\}$. Now if $f(A, B) = 0$ then $d(a, B) = 0$ for all $a \in A$. This implies that $a \in \overline{B} \; \forall \; a \in A$ (see Exercise 1.5.4). Hence $A \subseteq \overline{B}$. Thus, f is also not a metric. But we get a fairly reasonable idea of the definition of the following function:

$$d_H(A, B) = \max \left\{ \sup_{a \in A} d(a, B), \; \sup_{b \in B} d(b, A) \right\}.$$

Now we leave it as an exercise for the readers to prove that d_H satisfies the properties of a metric on the collection of all non-empty closed and bounded subsets of a metric space (X, d) (Exercise 1.5.7). Note that the metric d_H is known as *Hausdorff metric* or *Hausdorff distance*.

In the previous section, we briefly talked about a (topological) structure which was weaker than that of a metric space. Now we end this chapter with a brief discussion of normed linear spaces which are stronger than metric spaces. Recall that the concept of a metric was parallel to that of distances in $\mathbb{R}$. The idea of norm is actually motivated by the length $|x|$ of a vector x. Consequently, we need a vector space to define norm on it, unlike metric which was defined on an arbitrary non-empty set.

First let us recall the precise definition of vector spaces.

Definition 1.5.5. Let X be a non-empty set with a function $+ : X \times X \to X$, and a function $\cdot \; : \mathbb{K} \times X \to X$, where $\mathbb{K} = \mathbb{R}$ or $\mathbb{C}$. Then $(X, +, \cdot)$ is called a **vector space** (or **linear space**) over $\mathbb{K}$ if the following conditions are satisfied for all x, y, $z \in X$ and α, $\beta \in \mathbb{K}$:

   (i) $x + y = y + x$,
   (ii) $(x + y) + z = x + (y + z)$,

(iii) there exists $0 \in X$ such that $x + 0 = x$,
(iv) there exists $-x \in X$ such that $x + (-x) = 0$,
 (v) $\alpha \cdot (x + y) = \alpha \cdot x + \alpha \cdot y$,
(vi) $(\alpha + \beta) \cdot x = \alpha \cdot x + \beta \cdot x$,
(vii) $\alpha \cdot (\beta \cdot x) = (\alpha\beta) \cdot x$,
(viii) $1 \cdot x = x$

Note that if $(X, +)$ satisfies conditions (i) to (iv), then it is called an *abelian group*. The operation $+$ is usually called *vector addition*, while $\cdot$ is known as *scalar multiplication*. The vector space $(X, +, \cdot)$ is said to be a **real vector space** if $\mathbb{K} = \mathbb{R}$, and a **complex vector space** if $\mathbb{K} = \mathbb{C}$.

Evidently, $\mathbb{R}^n$ and l^p are real vector spaces with respect to co-ordinatewise operations, whereas $\mathbb{C}^n$ is a complex vector space with respect to co-ordinatewise operations defined as follows: $x + y = (x_1 + y_1, \ldots, x_n + y_n)$, $\lambda x = (\lambda x_1, \ldots, \lambda x_n)$, where $x = (x_1, \ldots, x_n)$, $y = (y_1, \ldots, y_n)$ and $\lambda \in \mathbb{C}$.

Definition 1.5.6. A **norm** is a real-valued function defined on a vector space X over $\mathbb{K}$ ($\mathbb{K} = \mathbb{R}$ or $\mathbb{C}$), denoted by $\|x\|$, which satisfies the following properties:

 (i) $\|x\| \geq 0$,
(ii) $\|x\| = 0$ if and only if $x = 0$,
(iii) $\|\alpha x\| = |\alpha| \, \|x\|$,
(iv) $\|x + y\| \leq \|x\| + \|y\|$, for every x, $y \in X$ and $\alpha \in \mathbb{K}$.

Moreover, the pair $(X, \|\cdot\|)$ is called a **normed linear space** or simply a **normed space**.

Example 1.5.1. The Euclidean space $\mathbb{R}^n$ is a normed linear space with a norm defined as follows:

$$\|x\| = \sqrt{|x_1|^2 + \cdots + |x_n|^2}, \text{ where } x = (x_1, \ldots, x_n) \in \mathbb{R}^n.$$

Hence the norm is called the **Euclidean norm**.

Interestingly, on a normed linear space $(X, \|\cdot\|)$ one can naturally define a metric as follows:

$$d(x, y) = \|x - y\|, \text{ where } x, y \in X.$$

It is known as the **metric induced by the norm**. Thus, every normed linear space is a metric space but every metric on a vector space need not be obtained from a norm.

Example 1.5.2. Let X be a nonempty set, and let $B(X)$ be the collection of all bounded real-valued functions defined on X. Then $B(X)$ is a vector space over $\mathbb{R}$ under the usual addition of functions on X and the usual scalar multiplication of f in $B(X)$ by a scalar $\alpha \in \mathbb{R}$. On $B(X)$, we can define the **supremum norm** (also called **uniform norm**) as follows:

$$\|f\|_\infty = \sup_{x \in X} |f(x)| \quad \text{for } f \in B(X).$$

Because of the notation $\| \cdot \|_\infty$, this norm is also called the *infinity norm*. This norm induces the supremum metric D on $B(X)$:
$D(f,g) = \|f - g\|_\infty \ \forall \ f, \ g \in B(X)$.

Exercises

1.5.1. Show that in the Euclidean space $\mathbb{R}^n$, the closure of any open ball $B(x, r)$ is the closed ball $C(x, r)$. Give an example to show that in general, they need not coincide.

1.5.2. For the non-empty subsets A and B of $\mathbb{R}$, define $A + B = \{a + b : a \in A, \ b \in B\}$.

(a) If A and B are both closed in $\mathbb{R}$, is $A + B$ also closed in $\mathbb{R}$?
(b) Let A be closed in $\mathbb{R}$ and $x \in \mathbb{R}$. Is $x + A = \{x + a : a \in A\}$ closed in $\mathbb{R}$?

1.5.3. Let (X, d) be a metric space and (x_n) be a sequence in X converging to a point x in X. Show that the set $F = \{x_n : n \in \mathbb{N}\} \cup \{x\}$ is closed in (X, d). Is x an accumulation point of F?

1.5.4. Let (X, d) be a metric space, A and B be two non-empty subsets of X and $x \in X$. Show that

(a) $d(A, B) = \inf\{d(a, B) : a \in A\} = \inf\{d(b, A) : b \in B\}$
(b) $d(x, A) = 0 \Leftrightarrow x \in \overline{A}$, that is, $\overline{A} = \{x \in X : d(x, A) = 0\}$
(c) If $A \subseteq B$, then $d(x, B) \leq d(x, A) \leq d(x, B) + \text{diam}(B)$
(d) $d(x, A) \leq d(x, y) + d(y, A) \ \forall \ y \in X$

(e) $d(x,y) \le d(x,A) + \operatorname{diam}(A) + d(y,A) \quad \forall\, y \in X$
(f) $d(A,B) \le d(x,A) + d(x,B)$
(g) $\operatorname{diam}(A) = \operatorname{diam}(\overline{A})$
(h) If $\{A_i : i \in I\}$ is a family of non-empty subsets of X, then $d(x, \bigcup_{i \in I} A_i) = \inf\{d(x,A_i) : i \in I\}$. If $\bigcap_{i \in I} A_i \ne \emptyset$, then what can you say regarding $d(x, \bigcap_{i \in I} A_i)$?

1.5.5. Suppose S is a non-empty subset of $\mathbb{R}$. If S is bounded below, then $d(\inf S, S) = 0$.

1.5.6. Let A be a non-empty subset of (X,d). For $\epsilon > 0$, let $A^\epsilon = \bigcup_{a \in A} B(a, \epsilon)$.

(i) Show that $A^\epsilon = \{x \in X : d(x,A) < \epsilon\}$.
(ii) Prove that $\overline{A} = \bigcap_{\epsilon > 0} A^\epsilon$.
(iii) Give an example of a subset A of some metric space such that $A^2 \ne (A^1)^1$.

1.5.7. (i) Prove that the following function d_H is a metric on the family, $\mathrm{CLB}(X)$, of all non-empty closed and bounded subsets of a metric space (X,d):

$$d_H(A,B) = \max\left\{ \sup_{a \in A} d(a,B), \ \sup_{b \in B} d(b,A) \right\}.$$

(ii) Verify that $d_H(A,B) = \inf\{\epsilon > 0 : B \subseteq A^\epsilon \text{ and } A \subseteq B^\epsilon\}$.
(iii) Find $d_H(A,B)$, where $A = [0,2]$ and $B = \{0, (-1)/n : n \in \mathbb{N}\}$ are closed and bounded subsets of $\mathbb{R}$.
(iv) Suppose (X,d) is bounded. Then prove that $(\mathrm{CLB}(X), d_H)$ is also bounded.

1.5.8. In a metric space with discrete metric, what kind of sequences are convergent?

1.5.9. Prove that every convergent sequence in a metric space is Cauchy and bounded. What can you say about the boundedness of a Cauchy sequence?

1.5.10. Let $a = (a_n) \in l^p$ for $1 \le p < \infty$. Show that the sequence $(x^{(k)})$ in l^p, where $x^{(k)} = (a_1, a_2, \ldots, a_k, 0, 0, \ldots)$, converges to a in l^p. Give an example to show that it need not hold true in l^∞.

1.5.11. Consider the subset c_{oo} of l^∞ defined as $c_{oo} = \{(x_n) \in l^\infty : x_n = 0 \,\forall\, n \geq k \text{ for some } k \in \mathbb{N}\}$. Then show that c_{oo} is neither closed nor open in l^∞.

1.5.12. Prove that the set c and c_o are closed in l^∞, where c denotes the space of all convergent sequences, while c_0 is the space of all sequences converging to zero.

1.5.13. Let $\| \cdot \|_1$ and $\| \cdot \|_2$ be norms on a vector space X. Prove that $\max\{\| \cdot \|_1, \| \cdot \|_2\}$ and $\| \cdot \|_1 + \| \cdot \|_2$ are also norms on X.

1.5.14. Prove that a metric d induced by a norm on a normed linear space $(X, \| \cdot \|)$ satisfies the following properties:

$$d(x + a, y + a) = d(x, y),$$
$$d(\alpha x, \alpha y) = |\alpha| \, d(x, y)$$

for all x, y, $a \in X$ and every $\alpha \in \mathbb{K} = \mathbb{R}$ or $\mathbb{C}$. Thus, a metric induced by a norm is translation invariant. It should be noted that if a metric satisfies these properties then it need not be induced by a norm. But these properties could be useful in proving some metric to be not induced by any norm.

1.5.15. Recall that a subset A of a normed linear space $(X, \| \cdot \|)$ (over $\mathbb{R}$) is called *convex* provided whenever $a_1, a_2 \in A$ and $\alpha \in [0, 1]$ then $\alpha a_1 + (1 - \alpha)a_2 \in A$, that is, for all a_1, $a_2 \in A$, the line segment joining these points, $\{(1 - \alpha)a_1 + \alpha a_2 : \alpha \in [0, 1]\}$, is also contained in A. Prove the following assertions:

(i) every open ball in $(X, \| \cdot \|)$ is convex.
(ii) if A is convex, then so is A^ϵ for $\epsilon > 0$.
(iii) the intersection of any family of convex subsets of X is again convex, and conclude that any subset of X has a smallest convex superset, that is, for $A \subseteq X$, there exists a smallest convex subset B of X such that $A \subseteq B$.
(iv) the sum of two convex sets is also convex.
(v) the interior and closure of a convex set are also convex.
(vi) a convex set is stable under taking convex combinations of its elements, that is, if $a_1, a_2, \ldots, a_n$ belong to a convex set A and $\alpha_1, \alpha_2, \ldots, \alpha_n$ are non-negative numbers such that $\alpha_1 + \alpha_2 + \ldots + \alpha_n = 1$ then $\alpha_1 a_1 + \alpha_2 a_2 + \ldots + \alpha_n a_n \in A$.

Note that a convex set can be defined in general for any vector space over $\mathbb{R}$.

1.6 Convergence of Subsequences

It is known that if a sequence (x_n) in a metric space (X, d) converges to some $x \in \mathbb{R}$, then any subsequence of (x_n) also converges to x. So now the question is: If a sequence (x_n) does not converge in a metric space, then what about its subsequences other than (x_n) itself? Do they converge? More precisely, which subsequences converge? Note that in this section, this question is answered for sequences in $\mathbb{R}$. And for that we need to talk about cluster points of a sequence. But we define such points in a general metric space.

Definition 1.6.1. Let (x_n) be a sequence in a metric space (X, d) and $x \in X$. Then x is called a **cluster point** of (x_n) in (X, d) if given $\epsilon > 0$ and $n \in \mathbb{N}$, there exists $k > n$ (k depends on both n and ϵ) such that $d(x_k, x) < \epsilon$.

Remark. Some authors call cluster points by 'limit points'. But in order to avoid ambiguity between 'limit point' and 'limit' of a sequence, we do not prefer to use the term 'limit point'. In fact, the term 'cluster point' is suitable in literal sense as well: keep $\epsilon > 0$ fixed but keep varying $n \in \mathbb{N}$, then we can find infinitely many members of the sequence in the ϵ-nhood of a cluster point.

Observe that if $x_n \to x$ in (X, d), then the sequential terms get arbitrarily close to the limit x of the sequence *eventually*. But the sequential terms get arbitrarily close to a cluster point *frequently*. Unlike limit of a sequence, there could be more than one cluster point of a sequence. For example, consider the sequence (x_n) where $x_{2n-1} = 1/n$ and $x_{2n} = 1 + 1/n$ for $n \in \mathbb{N}$. Then 0 and 1 are the cluster points of (x_n). In this example, note that the subsequence (x_{2n-1}) converges to 0, while the subsequence (x_{2n}) converges to 1. Subsequently, we have the following equivalent characterization of a cluster point of a sequence.

Theorem 1.6.1. *Let (x_n) be a sequence in a metric space (X, d) and $x \in X$. Then x is a cluster point of (x_n) in (X, d) if and only* *if there exists a subsequence (x_{k_n}) of (x_n) such that $x_{k_n} \to x$ in (X, d).*

Proof. First suppose that x is a cluster point of (x_n). For $\epsilon = 1$, choose $k_1 \in \mathbb{N}$ such that $d(x_{k_1}, x) < 1$. Now again apply the definition of cluster point for $\epsilon = \frac{1}{2}$ and $k_1 \in \mathbb{N}$, then there exists $k_2 \in \mathbb{N}$ such that $k_2 > k_1$ and $d(x_{k_2}, x) < \frac{1}{2}$. By induction, suppose $k_1, k_2, \ldots, k_n$ have been selected such that $k_1 < k_2 < \cdots < k_n$ and $d(x_{k_i}, x) < \frac{1}{i}$, $1 \le i \le n$. Then for $\epsilon = \frac{1}{n+1}$ and k_n, we may choose by the definition of a cluster point, $k_{n+1} \in \mathbb{N}$ such that $k_{n+1} > k_n$ and $d(x_{k_{n+1}}, x) < \frac{1}{n+1}$. Consequently, (x_{k_n}) is a subsequence of (x_n) such that $x_{k_n} \to x$.

Conversely, suppose $x_{k_n} \to x$. Let $m \in \mathbb{N}$ and $\epsilon > 0$. Since $x_{k_n} \to x$, there exists $n_o \in \mathbb{N}$ such that $d(x_{k_n}, x) < \epsilon \; \forall \, n \ge n_o$. Now choose $n_1 > \max\{n_o, m\}$. Then $d(x_{k_{n_1}}, x) < \epsilon$ and also $k_{n_1} \ge n_1 > m$. Thus, x is a cluster point of (x_n). $\qquad\square$

It is evident that a sequence in a metric space need not have a cluster point. Consider the simple example of the sequence of natural numbers. Note that this sequence is unbounded. Now the question is: *Can you construct a bounded sequence in $\mathbb{R}$ which has no cluster point?* The answer is negative! Interestingly, every bounded sequence in $\mathbb{R}$ has a convergent subsequence. This is given by the well-known result, *Bolzano–Weierstrass theorem*. For proving this theorem, we need to do some more analysis. So let us begin.

Definitions 1.6.2. A sequence (x_n) in $\mathbb{R}$ is said to be

(a) **increasing** if $x_n \le x_{n+1} \; \forall \, n \in \mathbb{N}$.
(b) **decreasing** if $x_n \ge x_{n+1} \; \forall \, n \in \mathbb{N}$.
(c) **monotone** if (x_n) is either increasing or decreasing.

Evidently, not every sequence in $\mathbb{R}$ is monotone but every sequence in $\mathbb{R}$ has a monotone subsequence (Exercise 1.6.3). Now we look at the condition under which a monotone sequence is convergent. Definitely, for the sequence to be convergent, it should be bounded first. But is boundedness sufficient for the convergence of a monotone sequence in $\mathbb{R}$? The answer is affirmative.

Theorem 1.6.2 (Monotone convergence theorem). *A monotone sequence of real numbers is convergent if it is bounded.*

Proof. Suppose (x_n) is a bounded, increasing sequence in $\mathbb{R}$. Then there exists $M > 0$ such that $x_n \leq M \ \forall \ n \in \mathbb{N}$. By completeness axiom on $\mathbb{R}$, $\sup\{x_n : n \in \mathbb{N}\}$ exists in $\mathbb{R}$. Let $x^* = \sup\{x_n : n \in \mathbb{N}\}$. We claim that x^* is the limit of the sequence (x_n).

Let $\epsilon > 0$. By Theorem 1.2.1, there exists $k \in \mathbb{N}$ such that

$$x^* - \epsilon < x_k \leq x_n \leq x^* < x^* + \epsilon \ \forall \ n \geq k.$$

Thus,

$$|x_n - x^*| < \epsilon \ \forall \ n \geq k.$$

Hence $x_n \to x^*$.

For proving a bounded, decreasing sequence (y_n) to be convergent, note that the sequence $(-y_n)$ is bounded and increasing. $\qquad \square$

Hence monotonicity establishes a link between boundedness and convergence of a real sequence. But a bounded sequence in $\mathbb{R}$ need not be monotone in general. So now the next step is to construct somehow a monotone sequence out of a given bounded sequence in $\mathbb{R}$.

Let (x_n) be a bounded sequence in $\mathbb{R}$. For each $n \in \mathbb{N}$, let

$$v_n = \sup_{k \geq n} x_k \quad \text{and} \quad w_n = \inf_{k \geq n} x_k.$$

Since (x_n) is bounded, (v_n) and (w_n) are bounded sequences of real numbers. Moreover, note that the sequence (v_n) is decreasing, while (w_n) is an increasing sequence in $\mathbb{R}$. Hence by monotone convergence theorem,

$$\lim_{n \to \infty} v_n = \inf_{n \in \mathbb{N}} v_n = \inf_{n \in \mathbb{N}} \left[\sup_{k \geq n} x_k \right] \quad \text{and}$$

$$\lim_{n \to \infty} w_n = \sup_{n \in \mathbb{N}} w_n = \sup_{n \in \mathbb{N}} \left[\inf_{k \geq n} x_k \right].$$

These limits are given special names as they play a crucial role in real analysis.

Definitions 1.6.3. Let (x_n) be a bounded sequence in $\mathbb{R}$. Then

(i) the **limit superior** of (x_n) is defined by

$$\limsup x_n = \inf_{n \in \mathbb{N}} \left[\sup_{k \geq n} x_k \right].$$

(ii) the **limit inferior** of (x_n) is defined by

$$\liminf x_n = \sup_{n \in \mathbb{N}} \left[\inf_{k \geq n} x_k \right].$$

The next result gives a convenient way to find the limit superior and the limit inferior of a bounded sequence.

Theorem 1.6.3. *Let (x_n) be a bounded sequence in $\mathbb{R}$. Then $\liminf x_n$ and $\limsup x_n$ are the smallest and the largest cluster points of (x_n).*

Proof. Suppose (x_n) is a bounded sequence in $\mathbb{R}$. Let $s = \limsup x_n = \inf_{n \in \mathbb{N}} v_n$, where $v_n = \sup_{k \geq n} x_k$. We first show that s is a cluster point of (x_n). Let $\epsilon > 0$ and $m \in \mathbb{N}$. Recall that (v_n) is a decreasing sequence which converges to s. Hence there exists $n > m$ such that $s \leq v_n < s + \epsilon$. Since $v_n = \sup_{k \geq n} x_k$, $x_k < s + \epsilon \ \forall \ k \geq n$. On the other hand, there must exist some $k \geq n$ such that $s - \epsilon < x_k$. If not, then $x_k \leq s - \epsilon \ \forall \ k \geq n$ implying $v_n \leq s - \epsilon$. But $s \leq v_n$. Hence there exists $k \geq n$ such that $s - \epsilon < x_k < s + \epsilon$. But $n > m$ and so $k > m$. Hence s is a cluster point of (x_n).

Now we show that s is the largest cluster point of (x_n). Let x be a cluster point of (x_n) and choose $\epsilon > 0$. Then given $n \in \mathbb{N}$, $\exists \ m > n$ such that $x - \epsilon < x_m < x + \epsilon$. Then $x - \epsilon < x_m \leq \sup_{k \geq n} x_k$. Thus $x - \epsilon < v_n$ for all $n \in \mathbb{N}$. Hence $x - \epsilon \leq \inf_{n \in \mathbb{N}} v_n = s$, that is, $x - \epsilon \leq s \ \forall \ \epsilon > 0$. Therefore, $x \leq s$ and we are done.

Similarly, it can be proved that $\liminf x_n$ is the smallest cluster point of (x_n). $\qquad\square$

Remark. As a consequence, we have

$$\liminf x_n \leq \limsup x_n.$$

Now the question is: *When does the reverse inequality hold true?* (see Exercise 1.6.4.)

Note that Theorem 1.6.3 also talks about the existence of a cluster point of a bounded sequence in $\mathbb{R}$. Hence, every bounded sequence in $\mathbb{R}$ has a cluster point. Then by Theorem 1.6.1, it can be concluded that every bounded sequence in $\mathbb{R}$ has a convergent subsequence. Hence we have finally proved the next result.

Theorem 1.6.4 (Bolzano–Weierstrass theorem). *Every bounded sequence in $\mathbb{R}$ has a convergent subsequence.*

Remark. Note that we do not have an analogous version of Bolzano–Weierstrass Theorem for any general metric space. For example, consider the l^2-space. Then the sequence (e_n), where $e_n \in l^2$ with 1 at the nth place and 0 otherwise, is bounded but (e_n) has no convergent subsequence.

Exercises

1.6.1. Suppose (k_n) and (m_n) are strictly increasing sequences of natural numbers such that $\{k_n, m_n : n \in \mathbb{N}\} = \mathbb{N}$. Show that a sequence (x_n) in $\mathbb{R}$ converges if and only if both subsequences (x_{k_n}) and (x_{m_n}) of (x_n) converges to the same limit.

1.6.2. (a) How many cluster points does a convergent sequence have?

(b) If a bounded sequence (x_n) in a metric space (X, d) has a unique cluster point x, does this imply that $x_n \to x$? What can you say if $(X, d) = (\mathbb{R}, |\cdot|)$?

1.6.3. Let (x_n) be a sequence in $\mathbb{R}$. Prove that there exists a subsequence of (x_n) which is monotone. Using this observation, can you give an alternate proof for Bolzano–Weierstrass Theorem?

1.6.4. If (x_n) is a bounded sequence in $\mathbb{R}$, then show that $x_n \to x$ if and only if $\liminf x_n = \limsup x_n = x$.

1.6.5. Let (x_n) and (y_n) be two bounded sequences in $\mathbb{R}$. Prove that:

(i) $\limsup(-x_n) = -\liminf x_n$
(ii) $\liminf(-x_n) = -\limsup x_n$
(iii) $\limsup(x_n + y_n) \leq \limsup x_n + \limsup y_n$
(iv) $\liminf(x_n + y_n) \geq \liminf x_n + \liminf y_n$.

1.6.6. Show that the set of all cluster points of a sequence in (X, d) is closed in X.

1.6.7. Let (x_n) be a sequence in (X, d). Show that each accumulation point of the set $\{x_n : n \in \mathbb{N}\}$ is a cluster point of (x_n). What can you say about the converse?

1.6.8. Give an example of a sequence in $\mathbb{R}$ whose set of cluster points is $\mathbb{R}$. Generalize this construction to an arbitrary separable metric space.

Chapter 2

Continuity and Some Stronger Notions

We all are familiar with the significance of continuous functions from calculus and real analysis. In this chapter, we talk about continuity of a function whose domain and codomain are general metric spaces. Moreover, some very interesting classes of functions are also considered which lie strictly inside the class of continuous functions. These classes of special functions play a crucial role in various branches of mathematics.

2.1 Basics of Continuity

In this section, we mainly deal with the movement from one metric space to another. But here the following question arises: Why should one bother about moving from one metric space to another one? Let us try to answer this question. Suppose you have two metric spaces (X, d) and (Y, ρ). While the space (X, d) seems familiar to you, the space (Y, ρ) you may not know well of. Is it possible to have information on (Y, ρ) through (X, d) and vice-versa? For example, suppose that you want to know if (Y, ρ) is separable, while you know that (X, d) is separable. Actually, like a good vehicle, we should be able to drive both ways: from (X, d) to (Y, ρ) and from (Y, ρ) to (X, d). That means we should have information on the vehicle itself which we use. So one should expect that the nature of the vehicle will depend on the kind of information we try to gather regarding (Y, ρ) or (X, d).

Recall that a natural way to move from a set X to another set Y is to use a function from X to Y. But when we have metric spaces (X, d) and (Y, ρ), then the function from X to Y should have something more so that we can use the metric structures of X and Y profitably. This 'something more' actually depends on the extent to which we need the metric structures on X and Y to be related. We start with a very basic map which is usually found in abundance. These are known as *continuous maps*. Pay attention on the adjective 'continuous'. Intuitively, it means a map f from (X, d) to (Y, ρ) such that we move continuously: if we move slowly and methodically from x_1 to x_2 in X, then the movement from $f(x_1)$ to $f(x_2)$ is also slow and methodical. This intuition leads us to define continuity at a point x in X. Recall the ϵ–δ definition of continuity of a function $f : \mathbb{R} \to \mathbb{R}$. Here obviously we consider the usual distance metric $|\cdot|$. Let $f : (\mathbb{R}, |\cdot|) \to (\mathbb{R}, |\cdot|)$ be a map and $x_o \in X$. Then f is said to be *continuous* at x_o if given $\epsilon > 0$, there exists a $\delta > 0$ (δ depends on ϵ and x_o) such that $|f(x) - f(x_o)| < \epsilon$ whenever $|x - x_o| < \delta$. Let us imitate this definition for a map $f : (X, d) \to (Y, \rho)$ between two metric spaces.

Definition 2.1.1. Let $f : (X, d) \to (Y, \rho)$ be a map between two metric spaces (X, d) and (Y, ρ) and $x_o \in X$. Then f is said to be **continuous** at x_o if for every $\epsilon > 0$, there exists a $\delta > 0$ (δ depends on ϵ and x_o) such that $\rho(f(x), f(x_o)) < \epsilon$ whenever $d(x, x_o) < \delta$.

In other words, $f(x)$ can be made *arbitrarily* close to $f(x_o)$ provided x is *sufficiently* close to x_o. Here the reader should carefully observe the usage of the words in order to remove the confusion between ϵ and δ. First decide how much close to $f(x_o)$ you want the images of points (under f) to be? That measurement is given by ϵ. Then in order to ensure this ϵ-closeness, we need to decide how much close to x_o the pre-images should be? The second measurement is given by δ. Obviously, δ will depend on ϵ and x_o.

Now let us go back to Definition 2.1.1. Note that $d(x, x_o) < \delta$ means $x \in B_X(x_o, \delta)$ (the open ball centred at x_o with radius δ in (X, d)). Similarly $\rho(f(x), f(x_o)) < \epsilon$ means $f(x) \in B_Y(f(x_o), \epsilon)$. We are distinguishing between the open balls in (X, d) and those in (Y, ρ) by using the subscripts X and Y. Since an open ball centred at a point denotes a nhood of the point, Definition 2.1.1 can be rewritten in terms of nhoods as follows: the function $f : (X, d) \to (Y, \rho)$ is said to be continuous at x_o if given a nhood $B_Y(f(x_o), \epsilon)$ of $f(x_o)$ in Y, there exists a nhood $B_X(x_o, \delta)$ of x_o in X such that $f(B_X(x_o, \delta)) \subseteq B_Y(f(x_o), \epsilon)$.

Example 2.1.1. Let $f : \mathbb{R}^2 \setminus \{(0,0)\} \to \mathbb{R}$ be defined by

$$f(x, y) = \frac{x^2 - y^2}{x^2 + y^2} \text{ for } (x, y) \in \mathbb{R}^2 \setminus \{(0,0)\}.$$

Can you define $f(0,0)$ in such a way that f becomes continuous on $\mathbb{R}^2$? Let us think about the possible value of the function at $(0,0)$. Let $\epsilon = \frac{1}{2}$. We know that every open ball around $(0,0)$ contains points of the form (a, a) for suitable $a \in \mathbb{R}$. For $a \neq 0$, $f(a, a) = 0$. Hence by the continuity condition, $|f(0,0) - 0| < \frac{1}{2}$. This implies that $-\frac{1}{2} < f(0,0) < \frac{1}{2}$. Similarly, we can see that every nhood of $(0,0)$ contains points of x-axis. Thus, $|f(0,0) - f(a, 0)| = |f(0,0) - 1| < \frac{1}{2}$. Consequently, $\frac{1}{2} < f(0,0) < \frac{3}{2}$. A contradiction! Hence, f is not continuous at $(0,0)$ in any case.

Definition 2.1.2. A function $f : (X, d) \to (Y, \rho)$ between two metric spaces is said to be **continuous** on X (or simply continuous) if f is continuous at each point of X.

Here are some examples.

Example 2.1.2. Let $f : (X, d) \to (X, d)$ be the identity map, that is, $f(x) = x \ \forall \ x \in X$. For each point of X, given $\epsilon > 0$, we can choose $\delta = \epsilon$. So f is continuous at each point of X, that is, f is continuous on X.

Note that the identity map from $(\mathbb{R}, | \cdot |)$ to $(\mathbb{R}, d)$, where d is the discrete metric, is not continuous at any point of $\mathbb{R}$. Prove it! *What can you say regarding a function from $(\mathbb{R}, d)$ to $(\mathbb{R}, | \cdot |)$?*

Example 2.1.3. Let $f : (X, d) \to (Y, \rho)$ be a map between two metric spaces and $y_o \in Y$. If $f(x) = y_o \ \forall \ x \in X$, then f is continuous on X. In other words, a constant map is always continuous. Given $\epsilon > 0$, what is your δ?

Example 2.1.4. Let $f : (X, d) \to (Y, \rho)$ be a map such that $\rho(f(x), f(x')) = d(x, x') \ \forall \ x, \ x' \in X$. Such a map is called an *isometry.* The reader should verify that every isometry is continuous and one-to-one. But an isometry need not be onto. *Does there exist an isometry from* $(\mathbb{R}, |\cdot|)$ *onto* $(\mathbb{C}, |\cdot|)$? We will have a detailed discussion on isometries in the next section.

The next theorem presents the most useful characterizations of continuity.

Theorem 2.1.1. *Let* $f : (X, d) \to (Y, \rho)$ *be a function between two metric spaces. Then the following statements are equivalent:*

(a) f *is continuous on* X.
(b) $f^{-1}(U)$ *is open in* (X, d), *whenever* U *is open in* (Y, ρ).
(c) $f^{-1}(B_\rho(y, \epsilon))$ *is open in* (X, d) *for all* $y \in Y$ *and* $\epsilon > 0$.
(d) *Whenever* $x_n \to x$ *in* (X, d), *then* $f(x_n) \to f(x)$ *in* (Y, ρ) *for any* $x \in X$.
(e) $f(\overline{A}) \subseteq \overline{f(A)}$ *holds for every subset* A *of* X.
(f) $f^{-1}(C)$ *is closed in* (X, d), *whenever* C *is closed in* (Y, ρ).

Proof. (a) $\Rightarrow$ (b): Let U be open in (Y, ρ). In order to show that $f^{-1}(U)$ is open in (X, d), we will show that each point of $f^{-1}(U)$ is an interior point in (X, d).

Let $x \in f^{-1}(U)$. So $f(x) \in U$. Since U is open in (Y, ρ), there exists $\epsilon > 0$ such that $f(x) \in B_\rho(f(x), \epsilon) \subseteq U$. Since f is continuous at x, there exists $\delta > 0$ such that $f(B_d(x, \delta)) \subseteq B_\rho(f(x), \epsilon)$. So $x \in B_d(x, \delta) \subseteq f^{-1}(B_\rho(f(x), \epsilon)) \subseteq f^{-1}(U)$. Hence x is an interior point of $f^{-1}(U)$ and consequently $f^{-1}(U)$ is open in (X, d).

(b) $\Rightarrow$ (c): This is immediate.

(c) $\Rightarrow$ (d): Let (x_n) be a sequence in X converging to a point x in (X, d). We need to show that $f(x_n) \to f(x)$ in (Y, ρ), that is, we need to show that every nhood of $f(x)$ in (Y, ρ) eventually contains all the members of the sequence $(f(x_n))$.

Let U be a nhood of $f(x)$. Then there exists an $\epsilon > 0$ such that $f(x) \in B_\rho(f(x), \epsilon) \subseteq U$. By (c), $f^{-1}(B_\rho(f(x), \epsilon))$ is open in (X, d). Moreover, it contains x. Hence there exists $\delta > 0$ such that $x \in B_d(x, \delta) \subseteq f^{-1}(B_\rho(f(x), \epsilon))$. Since $x_n \to x$ in (X, d), there exists $n_o \in \mathbb{N}$ such that $x_n \in B_d(x, \delta) \ \forall \ n \geq n_o$, that is, $x_n \in f^{-1}(B_\rho(f(x), \epsilon)) \ \forall \ n \geq n_o$. So $f(x_n) \in B_\rho(f(x), \epsilon) \ \forall \ n \geq n_o$. But that precisely means that $f(x_n) \to f(x)$ in (Y, ρ).

(d) $\Rightarrow$ (e): Let $x \in \overline{A}$ so that $f(x) \in f(\overline{A})$. By Theorem 1.5.2, there exists a sequence (x_n) in A such that $x_n \to x$ in (X, d). By (d), $f(x_n) \to f(x)$ in (Y, ρ). But $x_n \in A \Rightarrow f(x_n) \in f(A)$. Hence again by Theorem 1.5.2, we get, $f(x) \in \overline{f(A)}$. Thus, $f(\overline{A}) \subseteq \overline{f(A)}$.

(e) $\Rightarrow$ (f): Let C be a closed set in (Y, ρ). Hence $C = \overline{C}$ in (Y, ρ). Let $A = f^{-1}(C)$, then $f(A) \subseteq C$. By (e), $f(\overline{A}) \subseteq \overline{f(A)} \subseteq \overline{C} = C$. This implies that $\overline{A} \subseteq f^{-1}(C) = A$. Hence $\overline{A} = A$, that is, $A = f^{-1}(C)$ is closed in (X, d).

(f) $\Rightarrow$ (a): We need to show that f is continuous at every point of X. Let $x \in X$ and choose $\epsilon > 0$. Note that the set $C = Y - B_\rho(f(x), \epsilon) = \{y \in Y : \rho(f(x), y) \geq \epsilon\}$ is closed in (Y, ρ). Hence by (f), $f^{-1}(C)$ is closed in (X, d). Since $x \notin f^{-1}(C)$ and $X - f^{-1}(C)$ is open in (X, d), there exists $\delta > 0$ such that $x \in B_d(x, \delta) \subseteq X - f^{-1}(C)$. Now if $x' \in B_d(x, \delta)$, then $x' \in X - f^{-1}(C)$. This implies that $f(x') \in B_\rho(f(x), \epsilon)$. So we get $f(B_d(x, \delta)) \subseteq B_\rho(f(x), \epsilon)$. Hence f is continuous at x. $\square$

Remark. Let A be a non-empty subset of X. We call a function $f : (X, d) \to (Y, \rho)$ to be continuous on A if f is continuous at each point of A. It should not be confused with the continuity of the restriction of f on A, that is, $f : (A, d) \to (Y, \rho)$. For example, consider the function $f : \mathbb{R} \to \mathbb{R}$ defined as $f(x) = 0$ for $x \in \mathbb{Q}$ and $f(x) = 1$ for $x \notin \mathbb{Q}$. Then the restriction of f on $\mathbb{Q}$ is a constant function and hence it is continuous. But f is not continuous on $\mathbb{Q}$ because by the density of irrationals in $\mathbb{R}$ for every rational number a there exists a sequence (x_n) of irrationals converging to a, but $f(x_n) \nrightarrow f(a)$. In particular, take the sequence $(\frac{\sqrt{2}}{n})$ which converges to 0. The reader should observe that one side implication is always true: if f is continuous on A then the function $f : (A, d) \to (Y, \rho)$ is continuous (use Theorem 2.1.1(d)). The converse holds true if A is open: let $a \in A$ and (x_n) be a sequence in X which converges to a. Since A is open, the sequential terms of (x_n) are eventually contained in A and hence again by using the sequential characterization of continuity of $f : (A, d) \to (Y, \rho)$ we get that f is continuous at a.

Many a times, when a function is not continuous on the whole domain but on discrete points then in that scenario sequential characterization for continuity could be highly convenient. Hence now we manipulate the condition (d) of the previous theorem for continuity at discrete points of the domain space.

Theorem 2.1.2. *Let* $f : (X, d) \to (Y, \rho)$ *be a function between two metric spaces and* $x \in X$. *Then the following statements are equivalent:*

(a) *f is continuous at x.*
(b) *Whenever $x_n \to x$ in (X, d), then $f(x_n) \to f(x)$ in (Y, ρ).*

Proof. (a) $\Rightarrow$ (b): Suppose $x_n \to x$ in (X, d). To show that $f(x_n) \to f(x)$ in (Y, ρ), choose an open ball $B_\rho(f(x), \epsilon)$ around $f(x)$ in (Y, ρ). Since f is continuous at x, there exists an open ball $B_d(x, \delta)$ around x in (X, d) such that $f(B_d(x, \delta)) \subseteq B_\rho(f(x), \epsilon)$. Since $x_n \to x$ in (X, d), there exists an $n_\delta \in \mathbb{N}$ such that $x_n \in B_d(x, \delta)$ $\forall$ $n \geq n_\delta$. Hence $f(x_n) \in B_\rho(f(x), \epsilon)$ $\forall$ $n \geq n_\delta$, that is, $f(x_n) \to f(x)$ in (Y, ρ). (Note that n_δ depends on δ which in turn depends on ϵ, hence n_δ depends on ϵ.)

(b) $\Rightarrow$ (a): If possible, suppose that f is not continuous at x. Then there exists $\epsilon > 0$ for which no positive δ works. In particular, $\delta = \frac{1}{n}$ does not work for this ϵ. Hence there exists $x_n \in X$ such that $d(x, x_n) < \frac{1}{n}$, but $\rho(f(x), f(x_n)) \geq \epsilon$. This is true for all n in $\mathbb{N}$. Then $x_n \to x$ in (X, d), but $(f(x_n))$ does not converge to $f(x)$ in (Y, ρ). We get a contradiction. Hence f must be continuous at x. $\square$

Remarks. (i) Note that in (b) $\Leftrightarrow$ (a) of Theorem 2.1.2, we have implicitly applied the axiom of choice to get the existence of the sequence (x_n) with the required condition: for every $n \in \mathbb{N}$, consider the set $X_n = \{x' \in X : d(x, x') < \frac{1}{n}$ and $\rho(f(x), f(x')) \geq \epsilon\}$. Then each $X_n \neq \emptyset$ and hence by the axiom of choice (Proposition 1.1.2), $\prod_{n \in \mathbb{N}} X_n$ is non-empty. Let $(x_n) \in \prod_{n \in \mathbb{N}} X_n$. Then this sequence (x_n) converges to x but $(f(x_n))$ does not converge to $f(x)$.

(ii) Theorem 2.1.2 says that if f is continuous at x, then whenever $x_n \to x$, we have

$$\lim_{n \to \infty} f(x_n) = f(x) = f(\lim_{n \to \infty} x_n).$$

Note how we have interchanged the positions of $\lim_{n \to \infty}$ and f.

The next corollary is immediate from the previous result.

Corollary 2.1.1. *If the functions $f : (X, d_X) \to (Y, d_Y)$ and $g : (Y, d_Y) \to (Z, d_Z)$ are continuous, then their composition $g \circ f$ is also continuous.*

In calculus we repeatedly use the algebra of continuous functions. Now using the sequential characterization of continuity, we can prove similar results for real-valued continuous functions defined on *any* metric space.

Corollary 2.1.2. *Let f and g be real-valued functions defined on a metric space (X, d). Assume that both f and g are continuous at $x_o \in X$, then*

(a) *their sum $f + g$, defined as $(f + g)(x) = f(x) + g(x)$, is also continuous at x_o.*
(b) *the scalar multiple cf, defined as $(cf)(x) = cf(x)$, is also continuous at x_o, where $c \in \mathbb{R}$.*
(c) *their pointwise product fg, defined as $(fg)(x) = f(x)g(x)$, is also continuous at x_o.*
(d) *the reciprocal $\frac{1}{f}$, defined as $(\frac{1}{f})(x) = \frac{1}{f(x)}$, is also continuous at x_o, provided the reciprocal is defined, that is, $f(x) \neq 0$ on X.*

Remarks. (i) As a consequence of the previous result, one can see that $C(X, \mathbb{R})$, the set of all real-valued continuous functions on (X, d), form a vector space over $\mathbb{R}$ under usual addition and scalar multiplication.
(ii) In (d), the function f is not required to be non-zero on the whole X. If suppose $f(x_o) \neq 0$, then by the continuity of f, there exists a nhood around x_o such that $f(x) \neq 0$ on that nhood. Hence we can consider the reciprocal $\frac{1}{f}$ restricted to that nhood and similarly it can be proved that $\frac{1}{f}$ is continuous at x_o.

Using the previous result, one can easily verify that every polynomial function is also continuous on $\mathbb{R}$.

Exercises

2.1.1. Suppose $f : (X, d) \to (Y, \rho)$ is a function such that the inverse image of every closed ball in (Y, ρ) is closed in (X, d). Does this imply that f is continuous? Compare with Theorem 2.1.1(c).

2.1.2. Prove that a function $f : (X, d) \to (Y, \rho)$ is continuous if and only if it preserves convergent sequences, that is, if (x_n) is a convergent sequence in (X, d), then $(f(x_n))$ also converges in (Y, ρ). Compare it with Theorem 2.1.1 (d).

2.1.3. Let $f : (X, d) \to (Y, \rho)$ be a function. Then show that f is continuous if and only if $f^{-1}(B^\circ) \subseteq (f^{-1}(B))^\circ$ for every subset B of Y.

2.1.4. Prove that a function $f : (X, d) \to (Y, \rho)$ is continuous if and only if

$$d(x, A) = 0 \ \Rightarrow \ \rho(f(x), f(A)) = 0,$$

for all $x \in X$ and non-empty subsets A of X.

2.1.5. Prove that the function $x \mapsto \ln(2 + x^2)$ is continuous on $\mathbb{R}$.

2.1.6. Let f and g be two continuous functions from (X, d) into (Y, ρ). If there exists a dense subset A of X such that $f(x) = g(x)$ for all $x \in A$. Show that $f(x) = g(x)$ holds for all $x \in X$.

2.1.7. Prove that continuous image of a separable metric space is also separable.

2.1.8. Let (X, d) be a metric space in which every bounded sequence has a convergent subsequence. If $f : (X, d) \to (Y, \rho)$ is a continuous function, then show that f preserves boundedness, that is, $f(A)$ is bounded for every bounded subset A of (X, d). Recall that by Bolzano–Weierstrass theorem, every bounded sequence in $\mathbb{R}$ has a convergent subsequence. Hence, every continuous function on $\mathbb{R}$ preserves boundedness.

2.1.9. Let us define weaker notions of continuity. If f is a real-valued function on (X, d) and $x_o \in X$ then it is said to be *upper semi-continuous* at x_o if for every $\epsilon > 0$, there exists a $\delta > 0$ such that whenever $d(x, x_o) < \delta$, $f(x) < f(x_o) + \epsilon$. While f is called *lower semi-continuous* at x_o if for every $\epsilon > 0$, there exists a $\delta_1 > 0$ such that whenever $d(x, x_o) < \delta_1$, $f(x_o) - \epsilon < f(x)$. Clearly, a function is continuous at x_o if and only if it is both upper semi-continuous and lower semi-continuous at x_o.

(a) Give an example of a function which is upper (or lower) semi-continuous at a point but not continuous at the point.

(b) If $A \subseteq X$ and $f : (X, d) \to \mathbb{R}$ is a function defined as $f(x) = 1$ if $x \in A$ and $f(x) = 0$ if $x \notin A$, then show that f is lower semi-continuous at every point of X if and only if A is an open set. What about upper semi-continuity?

(c) Show that f is upper semi-continuous at x_o if and only if $-f$ is lower semi-continuous at x_o.

(d) Let (f_n) be a sequence of real-valued functions defined on (X, d) and let $f(x) = \inf_{n \in \mathbb{N}} f_n(x) \in \mathbb{R}$ for all $x \in X$. Show that if f_n is upper semi-continuous at $x_o \in X$ for every $n \in \mathbb{N}$, then the function f is also upper semi-continuous at x_o.

(e) Establish that f is upper semi-continuous at every point of X if and only if $f^{-1}(-\infty, a)$ is open in X for all $a \in \mathbb{R}$. And f is lower semi-continuous at every point of X if and only if $f^{-1}(a, \infty)$ is open in X for all $a \in \mathbb{R}$.

(f) Verify that f is upper semi-continuous at x_o if and only if whenever $x_n \to x_o$, then $\limsup f(x_n) \le f(x_o)$.

(g) Prove that f is upper semi-continuous at every point of X if and only if its hypograph $\{(x, t) \in X \times \mathbb{R} : t \le f(x)\}$ is closed in $(X \times \mathbb{R}, \mu)$, where μ is any of the metrics defined in Example 1.3.4.

(h) If f and g are upper semi-continuous at every point of X then show that the following functions are also upper semi-continuous at every point of X: $f + g$, $\lambda.f$ (where $\lambda \ge 0$), $\max\{f, g\}$ and $\min\{f, g\}$.

(i) Suppose that $f : (X, d) \to (Y, \rho)$ is continuous at $x_o \in X$ and $g : (Y, \rho) \to \mathbb{R}$ is upper semi-continuous at $f(x_o)$. Prove that their composition $g \circ f$ is upper semi-continuous at x_o.

2.1.10. ([22]) Let us call a function $f : (X, d) \to (Y, \rho)$ to be *sub-continuous* at x in X if for every sequence (x_n) in X converging to x, there exists a subsequence $(f(x_{k_n}))$ of $(f(x_n))$ converging to some y in Y. If f is subcontinuous at each x in X, then f is said to be subcontinuous. Clearly, every continuous function is subcontinuous.

(a) Give an example of a subcontinuous function which is not continuous.

(b) Show that a function $f : (X, d) \to (Y, \rho)$ is continuous if and only if it is subcontinuous and has closed graph, that is, the set $G(f) = \{(x, f(x)) : x \in X\}$ is closed in $(X \times Y, \mu_1)$ where $\mu_1(a, b) = d(x_1, x_2) + \rho(y_1, y_2)$ for $a = (x_1, y_1)$ and $b = (x_2, y_2)$.

2.2 Continuous Functions: New from Old

As we explore metric spaces further, we may require to produce new continuous functions with certain properties by using other continuous functions. So now we talk about some of the powerful results in this regard which do have their generalized versions for topological spaces but here we prove them primarily in the context of metric spaces. These results play a crucial role in the construction of new continuous functions in the upcoming chapters.

Theorem 2.2.1 (Urysohn's lemma). *Let A and B be non-empty disjoint closed subsets of a metric space (X, d). Then there exists a continuous function $f : (X, d) \to [0, 1]$ such that $f(x) = 0$ for all $x \in A$ and $f(x) = 1$ for all $x \in B$.*

Proof. Let us define a function $f : (X, d) \to \mathbb{R}$ as follows:

$$f(x) = \frac{d(x, A)}{d(x, A) + d(x, B)}.$$

First note that f is well-defined: if $d(x, A) + d(x, B) = 0$, then $x \in \overline{A} \cap \overline{B}$ (by Exercise 1.5.4). Since A and B are closed, $x \in A \cap B$, which is impossible. Note that $x \mapsto d(x, A)$ is a continuous function on (X, d) (in fact it satisfies much stronger property which is proved in Example 2.4.1 later). Hence by algebra of continuous functions f is also continuous. Moreover, it is clear that $0 \leq f(x) \leq 1$ on X. If $x \in A$ then $d(x, A) = 0$. This implies that $f(x) = 0$ on A and similarly, $f(x) = 1$ on B. $\square$

The Urysohn's Lemma says that disjoint closed sets in a metric space can be separated by a continuous function. Using this continuous function, one can prove that the disjoint closed sets can also be separated by open sets, that is, there exist disjoint open subsets U and V of (X, d) such that $A \subseteq U$ and $B \subseteq V$. Note that a topological space (X, τ) is said to be *normal* if every pair of disjoint closed subsets in (X, τ) can be separated by open sets. Normality is a very desirable property in topological spaces, and metric spaces possess this property. In our next corollary, we will prove that metric spaces are stronger than normal spaces.

Corollary 2.2.1. *Suppose that A and B are non-empty disjoint closed subsets of a metric space (X, d). Then there exist open subsets U and V in X such that $A \subseteq U$ and $B \subseteq V$ and $\overline{U} \cap \overline{V} = \emptyset$.*

Proof. By Urysohn's Lemma, there exists a continuous function $f : (X, d) \to [0, 1]$ such that $f(x) = 0$ for all $x \in A$ and $f(x) = 1$ for all $x \in B$. Let $U = f^{-1}[0, \frac{1}{3})$ and $V = f^{-1}(\frac{2}{3}, 1]$. Then by the continuity of f, U and V are open in X such that $A \subseteq U$ and $B \subseteq V$. Now we only need to prove that $\overline{U} \cap \overline{V} = \emptyset$. Suppose, if possible, $x \in \overline{U} \cap \overline{V}$. Then there exist (u_n) in U and (v_n) in V such that $u_n \to x$ and $v_n \to x$. Since f is continuous, $f(u_n) \to f(x)$ and $f(v_n) \to f(x)$. Now $u_n \in U \Rightarrow f(u_n) \in [0, \frac{1}{3})$. Similarly, $f(v_n) \in (\frac{2}{3}, 1]$. But then $f(x) \in [0, \frac{1}{3}] \cap [\frac{2}{3}, 1]$. A contradiction! $\qquad\square$

Theorem 2.2.2 (Pasting lemma or gluing lemma). *Let $\{A_i : i \in I\}$ be a non-empty family of non-empty open subsets of a metric space (X, d) such that $X = \bigcup_{i \in I} A_i$. For each $i \in I$, let $f_i : (A_i, d) \to (Y, \rho)$ be a continuous function such that $f_i(x) = f_j(x)$ for all $x \in A_i \cap A_j$ and $i, j \in I$. Then the function $f : (X, d) \to (Y, \rho)$ defined as: $f(x) = f_i(x)$ for $x \in A_i$ is continuous on X.*

Proof. We will use Theorem 2.1.1(b) to prove the continuity of f. Let U_Y be an open subset of (Y, ρ). Then $f_i^{-1}(U_Y)$ is open in A_i for all $i \in I$. Consequently, $f_i^{-1}(U_Y) = U_i \cap A_i$ where U_i is open in (X, d). Since A_i is open in X, $f_i^{-1}(U_Y)$ is open in X and hence $\bigcup_{i \in I} f_i^{-1}(U_Y)$ is also open in X. Now

$$f^{-1}(U_Y) = f^{-1}(U_Y) \cap X = f^{-1}(U_Y) \cap \left(\bigcup_{i \in I} A_i \right) = \bigcup_{i \in I} (f^{-1}(U_Y) \cap A_i)$$

$$= \bigcup_{i \in I} f_i^{-1}(U_Y).$$

Hence $f^{-1}(U_Y)$ is open in X which further implies that f is continuous. $\qquad\square$

Remark. On similar lines, one can easily prove that if we instead take $\{A_i : i \in I\}$ to be a finite collection of non-empty closed subsets of a metric space (X, d), then the function f is again continuous. Here we are using the fact that the finite union of closed subsets is also closed. But this may not hold true for an arbitrary collection of closed sets (see Exercise 2.2.1).

Theorem 2.2.3. *Let (X, d_X), (Y, d_Y) and (Z, d_Z) be metric spaces and $f : X \to Y \times Z$ be a function defined by $f(x) = (f_1(x), f_2(x))$, where $f_1 : X \to Y$ and $f_2 : X \to Z$. Then f is continuous if and only if the functions f_1 and f_2 are continuous. Here we are considering the following metric on $Y \times Z$: $\mu_1(a, b) = d_Y(y_1, y_2) + d_Z(z_1, z_2)$ for $a = (y_1, z_1)$ and $b = (y_2, z_2)$.*

Proof. First, suppose that f is continuous. Consider the two projection maps $\pi_1 : Y \times Z \to Y : \pi_1(y, z) = y$ and $\pi_2 : Y \times Z \to Z : \pi_2(y, z) = z$. Then using sequential characterization of continuity (Theorem 2.1.1(d)), it is easy to see that π_1 and π_2 are continuous. Now $f_1 = \pi_1 \circ f$ and $f_2 = \pi_2 \circ f$ and composition of continuous functions is also continuous (Corollary 2.1.1), hence f_1 and f_2 are continuous.

Conversely, let f_1 and f_2 be continuous functions. For proving the continuity of f, we will use Theorem 2.1.1(d). So let $x_n \to x$ in (X, d_X) as $n \to \infty$. Then by the continuity of f_1 and f_2, as $n \to \infty$ $f_1(x_n) \to f_1(x)$ and $f_2(x_n) \to f_2(x)$ in (Y, d_Y) and (Z, d_Z), respectively. Let $\epsilon > 0$. Thus, there exist $n_o \in \mathbb{N}$ such that $d_Y(f_1(x_n), f_1(x)) < \frac{\epsilon}{2}$ and $d_Z(f_2(x_n), f_2(x)) < \frac{\epsilon}{2}$ for all $n \geq n_o$. Consequently, $\mu_1(f(x_n), f(x)) < \epsilon$ for all $n \geq n_o$. Thus, $f(x_n) \to f(x)$ as $n \to \infty$. $\square$

Remarks. (i) The functions f_1 and f_2 are referred to as the *coordinate functions* of f.
(ii) The result remains true if we replace the metric μ_1 with μ_∞ or μ_2 as defined in Example 1.3.4 (because they all are equivalent metrics, will discuss later).

Exercises

2.2.1. Let us define a collection $\mathcal{A}$ of subsets of a metric space (X, d) to be *locally finite* if every $x \in X$ has a nhood which intersects only finitely many $A \in \mathcal{A}$.

(a) Show that the collection $\mathcal{A} = \{[n, n+1] : n \in \mathbb{N}\}$ is locally finite in $\mathbb{R}$, whereas $\{[-\frac{1}{n}, \frac{1}{n}] : n \in \mathbb{N}\}$ is not locally finite in $\mathbb{R}$.

(b) Prove that the union of a locally finite collection of closed sets in (X, d) is again closed in (X, d).

(c) Prove that Theorem 2.2.2 holds true if $\{A_i : i \in I\}$ is taken to be a locally finite collection of closed subsets of (X, d), instead of open subsets in (X, d).

(d) Give an example to show that Theorem 2.2.2 may fail to hold if

   (i) $\{A_i : i \in I\}$ is an arbitrary infinite collection of closed sets.

   (ii) $\{A_i : i \in I\}$ is a finite collection of arbitrary subsets of (X, d).

2.2.2. Let f and g be two real-valued continuous functions on (X, d). Show that the functions $h_1(x) = \min\{f(x), g(x)\}$ and $h_2(x) = \max\{f(x), g(x)\}$ are continuous on X.

2.2.3. Let $\{(X_n, d_n) : n \in \mathbb{N}\}$ be a family of metric spaces and $f : (X, \rho) \to (\prod_{i=1}^{\infty} X_i, d)$ be a function defined by $f(x) = (f_n(x))_{n \in \mathbb{N}}$, where $f_n : X \to X_n$. Then prove that f is continuous if and only if the function f_n is continuous for all $n \in \mathbb{N}$. Here the metric d is defined as in Example 1.3.10.

2.2.4. Let (X, d) be a metric space. Prove that $C(X)$ is a finite-dimensional vector space if and only if X is a finite set. This is a nice application of Urysohn's lemma in Linear Algebra.

2.3 Stronger Notions of Continuity

In set theory, we define bijective functions to rename the objects in a set as per our convenience. In metric spaces since we deal with a rich structure on a set, now the functions need to be much more than mere bijections. The equivalent characterizations of continuity (Theorem 2.1.1) motivates us for the following concept.

Definition 2.3.1. Let $f : (X, d) \to (Y, \rho)$ be a map between two metric spaces. Then f is said to be a **homeomorphism** between (X, d) and (Y, ρ) if f is a bijection such that both f and f^{-1} are continuous. Moreover, if such a homeomorphism exists, then the metric spaces (X, d) and (Y, ρ) are called **homeomorphic**.

Example 2.3.1. Every open interval is homeomorphic to $\mathbb{R}$: consider the function $f : (-\frac{\pi}{2}, \frac{\pi}{2}) \to \mathbb{R}$ defined by $f(x) = \tan x$. Then the function f is not just continuous but even differentiable. It is strictly increasing because its derivative $\sec^2 x$ is positive on $(-\frac{\pi}{2}, \frac{\pi}{2})$. Hence it is one-to-one. Moreover, its inverse $f^{-1}(x) = \tan^{-1} x$ is also continuous on $\mathbb{R}$. This proves that $\mathbb{R}$ is homeomorphic to $(-\frac{\pi}{2}, \frac{\pi}{2})$. Here note that we have considered the usual distance metric on both $\mathbb{R}$ and $(-\frac{\pi}{2}, \frac{\pi}{2})$. Furthermore, it can be proved that every open and bounded interval (a, b) is homeomorphic to $(-\frac{\pi}{2}, \frac{\pi}{2})$ (take a line segment joining the corresponding end points of the intervals, that is, $a \mapsto -\frac{\pi}{2}$ and $b \mapsto \frac{\pi}{2}$). Since homeomorphism is a transitive relation (see Corollary 2.1.1), we can say that $\mathbb{R}$ is homeomorphic to any interval of the form (a, b) where $a, b \in \mathbb{R}$. *Can you think of a homeomorphism between $(0, 1)$ and $(1, \infty)$?*

The previous example also shows that boundedness is not preserved by homeomorphism. For proving two spaces to be homeomorphic, we only need to produce a homeomorphism between them. But for establishing the fact that the two spaces are not homeomorphic, we need to look for certain topological properties (a property that is preserved under homeomorphism) being possessed by exactly one of the metric spaces. Compactness and connectedness are two such properties which we will talk about in detail in the upcoming chapters. Using them it can be easily proved that $\mathbb{R}$ cannot be homeomorphic to any interval of the form $[a, b]$, $[a, b)$, $(a, b]$, $[a, \infty)$, $(-\infty, b]$.

Let us now define another stronger form of continuity which is widely known as *uniform* continuity. It was first introduced for real-valued functions on Euclidean spaces by Eduard Heine in 1870. Note the adjective, 'uniform'. In the definition of continuity, δ not only depends on ϵ but it also depends on x. If it is possible to choose a $\delta > 0$ which is uniform for all $x \in X$ then we call the function to be uniformly continuous. The precise definition is as follows.

Definition 2.3.2. A function $f : (X, d) \to (Y, \rho)$ between two metric spaces (X, d) and (Y, ρ) is called **uniformly continuous** if given $\epsilon > 0$, there exists a $\delta > 0$ (δ depends only on ϵ) such that whenever $x, x' \in X$ and $d(x, x') < \delta$, we have $\rho(f(x), f(x')) < \epsilon$.

Remarks. (a) Unlike continuity, uniform continuity is not defined at a point but on a set. Hence continuity is a local concept but uniform continuity is a global concept.

(b) We call the function f to be uniformly continuous on $A \subseteq X$ if for every $\epsilon > 0$, there exists a $\delta > 0$ such that whenever a, $a' \in A$ and $d(a, a') < \delta$, we have $\rho(f(a), f(a')) < \epsilon$.

The geometrical difference between continuity and uniform continuity can be beautifully analyzed through the graph of the function $f : (0, \infty) \to \mathbb{R} : f(x) = \frac{1}{x}$. As $x \to 0+$, $f(x) \to +\infty$ and the δ used in the definition of continuity depends on the point x as well. Thus, intuitively the function is continuous but not uniformly continuous. We will give a more formal proof for this assertion in the upcoming text. *What will happen if we take the domain to be (a, ∞) for some $a > 0$?*

Example 2.3.2. The function $f : (\mathbb{R}, |\cdot|) \to (\mathbb{R}, |\cdot|)$ given by $f(x) = x^2$ is continuous (by Corollary 2.1.2). But it is not uniformly continuous: we need to show that there exists some $\epsilon > 0$ for which no δ works. Let us prove it for $\epsilon = 2$. If $\delta > 0$, then there exists $n_o \in \mathbb{N}$ such that $\frac{1}{n} < \delta \ \forall \ n \geq n_o$. Choose $x = n_o$ and $x' = n_o + \frac{1}{n_o}$. Then $|x - x'| = \frac{1}{n_o} < \delta$. But $|f(x) - f(x')| = 2 + \frac{1}{n_o^2} > 2$. Since δ was an arbitrary positive number, f is not uniformly continuous on $\mathbb{R}$. *Is the function uniformly continuous on a closed and bounded interval?* We will answer this in a more general sense in Theorem 4.3.7.

By looking at the definition of uniform continuity, one can easily realize that any function with domain equipped with the discrete metric is uniformly continuous (take $\delta < 1$). This observation could be used in the formulation of counter examples later.

Example 2.3.3. Consider the identity function from $(\mathbb{Q}, d)$ to $(\mathbb{Q}, |\cdot|)$, where d is the discrete metric. Then it is uniformly continuous. Moreover, $(\mathbb{Q}, d)$ is bounded but $(\mathbb{Q}, |\cdot|)$ is not. Hence uniform continuity does not preserve boundedness (compare with Exercise 2.1.8). Additionally, observe that $(\mathbb{Q}, d)$ is complete because the Cauchy sequences with respect to discrete metric are eventually constant. But $(\mathbb{Q}, |\cdot|)$ is not complete (think of a sequence of rationals which converges to an irrational number). Consequently, we can conclude that completeness is also not preserved by uniformly continuous functions.

Note that in Theorem 2.1.1, we got a sequential characterization of continuity. What about a sequential characterization of uniform continuity? Since convergent sequences characterizes continuous functions, one might guess that Cauchy sequences play the analogous role for uniformly continuous functions. But uniformly continuous functions are much stronger than those functions which map Cauchy sequences of the domain space to Cauchy sequences of the target space. Let us have a closer look at such functions.

Definition 2.3.3. A function $f : (X, d) \to (Y, \rho)$ between two metric spaces is said to be **Cauchy continuous** if $(f(x_n))$ is Cauchy in (Y, ρ) for every Cauchy sequence (x_n) in (X, d).

Note that a Cauchy-continuous function is also referred to as *Cauchy-sequentially regular* (or CS-regular for short) in the literature.

Proposition 2.3.1. *Let $f : (X, d) \to (Y, \rho)$ be a function.*

(a) *If f is Cauchy continuous then f is continuous.*
(b) *If f is uniformly continuous then f is Cauchy continuous.*

Proof. We only need to supply the proof for (a) because (b) follows from definitions. Let $x_n \to x$ in (X, d). Then the sequence $\langle x_1, x, x_2, x, \ldots \rangle$ is Cauchy. Since f is Cauchy continuous, the sequence $\langle f(x_1), f(x), f(x_2), f(x), \ldots \rangle$ is Cauchy in (Y, ρ). Thus, for all $\epsilon > 0$ there exists an $n_o \in \mathbb{N}$ such that $\rho(f(x_n), f(x)) < \epsilon \ \forall \, n \geq n_o$. This implies that $f(x_n) \to f(x)$. Hence f is continuous by Theorem 2.1.1. $\qquad \square$

Consequently, the class of Cauchy-continuous functions is intermediate between the class of continuous functions and that of uniformly continuous functions. In fact, the inclusions are strict. For example, $f : (0, \infty) \to \mathbb{R} : f(x) = \frac{1}{x}$ is continuous but not Cauchy-continuous (consider the sequence $(\frac{1}{n})$ in $(0, \infty)$).

Example 2.3.4. The function $f : \mathbb{R} \to \mathbb{R} : f(x) = x^2$ is Cauchy-continuous but not uniformly continuous. Let (x_n) be a Cauchy sequence in $\mathbb{R}$. We need to show that (x_n^2) is also Cauchy. Let $\epsilon > 0$.

Since (x_n) is Cauchy, there exists $n_o \in \mathbb{N}$ such that $|x_n - x_m| < 1$ for all n, $m \geq n_o$. Since for any x, $y \in \mathbb{R}$ $\mid |x| - |y| \mid \leq |x - y|$, we get $|x_n| < |x_{n_o}| + 1$ for $n \geq n_o$. Let us denote $|x_{n_o}| + 1$ by M. Now again using the Cauchyness of (x_n), we can find $n_1 > n_o$ such that $|x_n - x_m| < \frac{\epsilon}{2M}$ for all n, $m \geq n_1$. Thus,

$$|x_n^2 - x_m^2| = |x_n + x_m||x_n - x_m| < 2M.\frac{\epsilon}{2M} = \epsilon \ \forall \ n, \ m \geq n_1.$$

Example 2.3.5. Consider the real-valued function $f(x) = \sin(x^2)$ defined on $\mathbb{R}$. It is easy to see that the composition of two Cauchy-continuous functions is also Cauchy-continuous. Thus, f is Cauchy-continuous being the composition of the Cauchy-continuous functions $h(x) = x^2$ and $g(x) = \sin(x)$: by mean value theorem we know that $|\sin x - \sin y| \leq |x - y|$ for all x, $y \in \mathbb{R}$. Hence, the function g is in fact uniformly continuous (take $\delta = \epsilon$). *Is f uniformly continuous?*

So now we need to look for some other type of sequences which will characterize uniform continuity. In this regard, we next define asymptotic sequences.

Definition 2.3.4. Let (x_n) and (x'_n) be a pair of sequences in a metric space (X, d). Then they are called **asymptotic** if $d(x_n, x'_n) \to 0$ as $n \to \infty$, that is, given $\epsilon > 0$, there exists $n_\epsilon \in \mathbb{N}$ such that $d(x_n, x'_n) < \epsilon \ \forall \ n \geq n_\epsilon$. Such a pair is denoted by $(x_n) \asymp (x'_n)$.

In $(\mathbb{R}, |\cdot|)$, the sequences (n) and $(n + \frac{1}{n})$ are asymptotic, that is, $(n) \asymp (n + \frac{1}{n})$. It is easy to see that if two sequences in a metric space (X, d) converge to the same limit in X, then the sequences are asymptotic. The next result talks about the converse in some sense. The routine proof (using triangle's inequality) is left as an exercise.

Proposition 2.3.2. *Let (x_n) and (x'_n) be a pair of asymptotic sequences in a metric space (X, d).*

(a) *If one of them is Cauchy, so is the other one.*
(b) *If (x_n) converges to $x \in X$, then (x'_n) also converges to x.*

Let us now prove the main result.

Theorem 2.3.1. *A map $f : (X, d) \to (Y, \rho)$ between two metric spaces is uniformly continuous if and only if f sends every pair of asymptotic sequences in X to a pair of asymptotic sequences in (Y, ρ).*

Proof. Let f be uniformly continuous. If $\epsilon > 0$, then there exists a $\delta > 0$ such that $d(x, x') < \delta \Rightarrow \rho(f(x), f(x')) < \epsilon$. Suppose $(x_n) \asymp (x_n')$, then there exists $n_\delta \in \mathbb{N}$ such that $d(x_n, x_n') < \delta \ \forall \ n \geq n_\delta$. This implies that $\rho(f(x_n), f(x_n')) < \epsilon \ \forall \ n \geq n_\delta$, that is, $(f(x_n)) \asymp (f(x_n'))$. (Note that n_δ depends on δ and δ depends on ϵ. Hence n_δ depends on ϵ indirectly.)

Conversely, let us assume that f is not uniformly continuous. Then there exists an $\epsilon > 0$ for which no $\delta > 0$ works. In particular, $\delta = \frac{1}{n}$ does not work for any $n \in \mathbb{N}$. Hence there exist x_n, x_n' in X such that $d(x_n, x_n') < \frac{1}{n}$, but $\rho(f(x_n), f(x_n')) \geq \epsilon$. Clearly, $(x_n) \asymp (x_n')$, but $(f(x_n))$ is not asymptotic to $(f(x_n'))$. A contradiction. Hence f must be uniformly continuous. $\qquad\qquad\square$

Remark. It should be noted that the axiom of choice is implicitly applied in the converse part of Theorem 2.3.1: for every $n \in \mathbb{N}$, consider the set $V_n = \{(x, x') \in X \times X : d(x, x') < \frac{1}{n} \text{ and } \rho(f(x), f(x')) \geq \epsilon\}$. Then each $V_n \neq \emptyset$ and hence by the axiom of choice (Proposition 1.1.2), $\prod_{n \in \mathbb{N}} V_n$ is non-empty. Let $((x_n, x_n')) \in \prod_{n \in \mathbb{N}} V_n$. Then the sequences (x_n) and (x_n') are asymptotic but $(f(x_n))$ and $(f(x_n'))$ are not asymptotic.

Example 2.3.6. The function $f(x) = e^x$ is not uniformly continuous on $(\mathbb{R}, |\cdot|)$. Choose $x_n = \ln(n)$ and $x_n' = \ln(n) + \frac{1}{n}$. Then $(x_n) \asymp (x_n')$, but $|f(x_n) - f(x_n')| = n(e^{1/n} - 1) \to 1$ as $n \to \infty$.

Towards the end of this section, we think about an analogue of pasting lemma for uniform continuity. Surprisingly, it does not hold true in general for uniform continuity. For example, consider the subsets $A_1 = \{(x, \frac{1}{x}) : x > 0\}$ and $A_2 = \{(x, 0) : x > 0\}$ of $\mathbb{R}^2$. Then both A_1 and A_2 are closed in $X = A_1 \cup A_2$. Here X is equipped with the metric induced by the metric μ_1 on $\mathbb{R}^2$ (refer to Example 1.3.4). Now let us define two real-valued functions f_1 and f_2, on A_1 and A_2 respectively, as follows: $f_1(x, \frac{1}{x}) = x$ and $f_2(x, 0) = 2x$. Then clearly f_1 and f_2 are uniformly continuous. But the function $f : X \to \mathbb{R}$ defined as: $f(x, y) = f_1(x, y)$ if $(x, y) \in A_1$; $f(x, y) = f_2(x, y)$ if $(x, y) \in A_2$, is not uniformly continuous. Because the sequences $\langle (n, \frac{1}{n}) \rangle$ and $\langle (n, 0) \rangle$ are asymptotic but their respective images under f are not asymptotic. Here note that A_1 and A_2 are disjoint. *What can you say in case they intersect?*

Exercises

2.3.1. Show that $(0,1)$ and $(1,\infty)$ are homeomorphic.

2.3.2. Give an example to show that completeness is not preserved by homeomorphism.

2.3.3. The characterization of continuity given in Theorem 2.1.1(b) motivates for the following notion which could be helpful in dealing with homeomorphisms: a function $f : (X,d) \to (Y,\rho)$ is said to be an **open map** if $f(U)$ is open in (Y,ρ) for every open set U in (X,d). Analogously, we call a function f to be a **closed map** if $f(U)$ is closed in (Y,ρ) for every closed set U in (X,d).

(a) Show that if f is a bijection, then f^{-1} is continuous if and only if f is an open map if and only if f is a closed map.
(b) Give an example of an open map which is not closed.
(c) Give an example of a closed map which is not open.
(d) Give an example of a discontinuous function which is both open and closed.

2.3.4. Show that the function $f : (a,\infty) \to \mathbb{R}$ defined by $f(x) = \frac{1}{x}$ is uniformly continuous for $a > 0$.

2.3.5. Suppose $\mathcal{A}$ is a family of sequences in a metric space (X,d). Let us define a relation $\sim$ on $\mathcal{A}$ as follows: for (x_n) and (x_n') in $\mathcal{A}$,

$$(x_n) \sim (x_n') \iff (x_n) \text{ and } (x_n') \text{ are asymptotic.}$$

Show that $\sim$ is an equivalence relation.

2.3.6. Can you find a pair of asymptotic sequences in $(B(X),D)$, where X is a non-empty set and D is the supremum metric on $B(X)$?

2.3.7. Prove Proposition 2.3.2.

2.3.8 ([54]). Let us call two sequences (x_n) and (x_n') to be *uniformly asymptotic* in (X,d) if for every $\epsilon > 0$, there exists $n_\epsilon \in \mathbb{N}$ such that $d(x_n, x_m') < \epsilon$ for all $n,\ m \geq n_\epsilon$. We denote such sequences by $(x_n) \asymp_u (x_n')$. Prove the following assertions:

(a) If (x_n) is Cauchy then $(x_n) \asymp_u (x_n)$.
(b) If sequences (x_n) and (x_n') converges to the same limit, then $(x_n) \asymp_u (x_n')$.

(c) $(x_n) \asymp_u (x'_n) \Rightarrow (x_n) \asymp (x'_n)$.

(d) If $(x_n) \asymp_u (x'_n)$, then (x_n) is Cauchy.

(e) If (x_n) and (x'_n) are Cauchy sequences and $(x_n) \asymp (x'_n)$ then $(x_n) \asymp_u (x'_n)$.

(f) A function $f : (X, d) \to (Y, \rho)$ is Cauchy-continuous if and only if it preserves uniformly asymptotic sequences, that is, if $(x_n) \asymp_u (x'_n)$ in (X, d) then $(f(x_n)) \asymp_u (f(x'_n))$ in (Y, ρ).

2.3.9. Let $f : (X, d) \to (Y, \rho)$ be a continuous function. Show that f is Cauchy-continuous if (X, d) is complete. Hence the function $f : \mathbb{R} \to \mathbb{R} : f(x) = e^x$ is Cauchy-continuous.

2.3.10. If the functions $f : (X, d_X) \to (Y, d_Y)$ and $g : (Y, d_Y) \to (Z, d_Z)$ are Cauchy-continuous, then prove that their composition $g \circ f$ is also Cauchy-continuous.

2.3.11. Prove that the composition of two uniformly continuous functions $f : (X, d_X) \to (Y, d_Y)$ and $g : (Y, d_Y) \to (Z, d_Z)$ is also uniformly continuous.

2.3.12. Prove Theorem 2.2.3 for uniform continuity, that is, f is uniformly continuous if and only if the functions $f_1 : X \to Y$ and $f_2 : X \to Z$ are uniformly continuous.

2.3.13. If f and g are two real-valued Cauchy-continuous functions defined on (X, d), then prove that their pointwise product is also Cauchy-continuous. Do we have similar result for uniformly continuous functions?

2.3.14. Give an example of a real-valued function f which is uniformly continuous on $[n-1, n]$ for all $n \in \mathbb{N}$ but not so on $[0, \infty)$.

2.3.15. Using Theorem 2.3.1, show that the function $f(x) = \sin(x^2)$ is not uniformly continuous on $\mathbb{R}$. Here note that f is continuous and bounded.

2.4 Lipschitz Functions

In this section, we would like to throw some light on another version of continuity which is in fact much stronger than uniform continuity.

Definition 2.4.1. A function $f : (X, d) \to (Y, \rho)$ is called **Lipschitz** if there exists $M \geq 0$ such that

$$\rho(f(x), f(x')) \leq M d(x, x') \text{ for all } x, \; x' \in X.$$

Here the number M is called a *Lipschitz constant* for the function f (the function f is also called M-Lipschitz). In fact, any number greater than M is also a Lipschitz constant for the function f. Evidently, every Lipschitz function is uniformly continuous (take $\delta = \frac{\epsilon}{M}$). Lipschitz functions play a vital role in different branches of mathematics. In fact, recall that Lipschitz continuity helps in establishing the uniqueness of the solution for an initial value problem in Picard's theorem.

Example 2.4.1. Let A be a non-empty subset of a metric space (X, d). Then the real-valued function f defined on (X, d) as

$$f(x) = d(x, A) = \inf\{d(x, a) : a \in A\}$$

is Lipschitz: by triangle's inequality we have $d(x, a) \leq d(x, x') + d(x', a)$ for all $x, \; x' \in X$ and $a \in A$. Thus, $d(x, A) \leq d(x, x') + d(x', a)$, which implies that $d(x, A) - d(x, x') \leq d(x', a)$. Since this is true for all $a \in A$, by the definition of infimum $d(x, A) - d(x, x') \leq d(x', A)$. This implies that $d(x, A) - d(x', A) \leq d(x, x')$. Similarly, by interchanging the roles of x and x' we get $d(x', A) - d(x, A) \leq d(x, x')$. Combining the two inequalities we get, $|d(x, A) - d(x', A)| \leq d(x, x')$ for all $x, \; x' \in X$. Thus, f is Lipschitz with a Lipschitz constant 1.

Let us now observe something. Suppose $f : (X, d) \to (Y, \rho)$ is a Lipschitz function. Let

$$L(f) := \inf\{M \geq 0 : M \text{ is a Lipschitz constant for } f\}.$$

Then equivalently, $L(f)$ can also be expressed as

$$L(f) = \sup\left\{\frac{\rho(f(x), f(x'))}{d(x, x')} : x, \; x' \in X, \; x \neq x'\right\}.$$

Thus, $L(f)$ is the least Lipschitz constant for f.

Remark. Many authors call $L(f)$ to be the *Lipschitz norm*. But it should be noted that $L(f)$ does not define a norm on the family of all Lipschitz functions from (X, d) to (Y, ρ) because for constant functions $L(f) = 0$. See Exercise 2.4.11.

The next result gives a very convenient method for proving a function to be Lipschitz.

Proposition 2.4.1. *Let $f : [a, b] \to \mathbb{R}$ be a function which is continuous on $[a, b]$ and differentiable on (a, b). Then f' is bounded on (a, b) if and only if f is Lipschitz on $[a, b]$.*

Proof. Suppose there exists $M > 0$ such that $|f'(z)| \leq M \; \forall \; z \in (a, b)$ $- (\star)$. Now for any $x, \; y \in [a, b]$, $x < y$, apply mean value theorem on $[x, y]$. Thus, there exists $c \in (x, y)$ such that

$$\frac{f(x) - f(y)}{x - y} = f'(c).$$

This implies that $|\frac{f(x)-f(y)}{x-y}| = |f'(c)| \leq M$, by $(\star)$. Hence $|f(x) - f(y)| \leq M|x - y| \; \forall \; x, \; y \in [a, b]$.

Conversely, suppose f is Lipschitz on $[a, b]$ with Lipschitz constant M. For $x_o \in (a, b)$,

$$f'(x_o) = \lim_{x \to x_o} \frac{f(x) - f(x_o)}{x - x_o}.$$

Since $|f(x) - f(x_o)| \leq M|x - x_o| \; \forall \; x \in (a, b)$, $|f'(x_o)| \leq M$. As x_o was arbitrarily chosen in (a, b), f' is bounded on (a, b). $\qquad \square$

Remark. A result analogous to Proposition 2.4.1 can be proved for any interval in $\mathbb{R}$. See Exercise 2.4.7.

Now can you prove that the function $f(x) = x^2$ is Lipschitz (and hence uniformly continuous) on every bounded interval of $\mathbb{R}$?

If Lipschitz constant for a Lipschitz function, $f : X \to X$, lies in the open interval $(0, 1)$ then such a special class of functions becomes more useful and could be used to establish the existence of a unique fixed point. Before proving the main result, let us give the required definitions first.

Definition 2.4.2. Let (X, d) be a metric space. Then a function $f : (X, d) \to (X, d)$ is said to be a **contraction** (or contraction mapping) if there exists a number $M, 0 < M < 1$, such that

$$d(f(x), f(x')) \leq Md(x, x') \text{ for all } x, \; x' \in X.$$

Definition 2.4.3. Let $f : (X, d) \to (X, d)$ be a function. Then a point $a \in X$ is called a **fixed point** for f if $f(a) = a$.

Thus, the function $f(x) = x^3$ has three fixed points, namely 0, 1 and -1, on $\mathbb{R}$ whereas the function $f(x) = \sin(x)$ has a unique fixed point. But the function $f(x) = x + 1$ has no fixed point. The next result gives a sufficient condition for the existence of a unique fixed point of a function. The result is named after S. Banach who first stated it in 1922. Interestingly, it has wide applications in different branches of mathematics, especially in differential equations and numerical analysis.

Theorem 2.4.1 (Banach fixed point theorem or contraction mapping principle). *Let $f : (X, d) \to (X, d)$ be a contraction on a complete metric space (X, d). Then f has a unique fixed point in X.*

Proof. Let $x_o \in X$. Define a sequence (x_n) iteratively as: $x_n = f(x_{n-1})$ for $n \in \mathbb{N}$. We will prove that the sequence (x_n) is Cauchy and the limit of the sequence is the required fixed point of f. First note that for $k \in \mathbb{N}$,

$$d(x_{k+1}, x_k) = d(f(x_k), f(x_{k-1}))$$
$$\leq Md(x_k, x_{k-1}) \leq M^2 d(x_{k-1}, x_{k-2})$$
$$\leq \cdots \leq M^k d(x_1, x_o),$$

where $M, 0 < M < 1$, is a Lipschitz constant for f. Now for $n, m \in \mathbb{N}$ with $n > m$, by the triangle's inequality we have

$$d(x_n, x_m) \leq d(x_n, x_{n-1}) + d(x_{n-1}, x_{n-2}) + \cdots + d(x_{m+1}, x_m)$$
$$\leq (M^{n-1} + M^{n-2} + \cdots + M^m)d(x_1, x_o)$$
$$\leq \frac{M^m}{1 - M}d(x_1, x_o).$$

Since $M < 1$, $M^m \to 0$ as $m \to \infty$. Thus, (x_n) is a Cauchy sequence in the complete metric space (X, d). Let $x_n \to x'$ in X. Now using the continuity of f, we get

$$f(x') = f(\lim_{n \to \infty} x_n) = \lim_{n \to \infty} f(x_n) = \lim_{n \to \infty} x_{n+1} = x'.$$

Consequently, x' is a fixed point of f.

Uniqueness: Let $x'' \in X$ be another fixed point for f. Now

$$d(x', x'') = d(f(x'), f(x'')) \leq Md(x', x'') < d(x', x'').$$

This gives a contradiction. Thus, $x' = x''$. $\square$

Remarks. (i) In the previous proof, observe that the fact that
$M < 1$ is used in proving both the existence and uniqueness of
the fixed point for the contraction.

(ii) If the hypothesis of completeness of the metric space (X, d) is
violated in the previous result, then the contraction may fail to
possess any fixed point in X: consider $f : (0, 1) \to (0, 1)$ defined
as $f(x) = \frac{x}{2}$. The only possible fixed point for the contraction f
is 0 but $0 \notin (0, 1)$.

(iii) If the contraction condition is replaced by $d(f(x), f(x')) <$
$d(x, x')$, then also the conclusion may fail to hold: let $f :$
$[0, \infty) \to [0, \infty) : f(x) = \sqrt{1 + x^2}$. Since $|f'(x)| < 1$ for
$x \in (0, \infty)$, by mean value theorem $|f(x) - f(x')| < |x - x'|$
for all x, $x' \in [0, \infty)$, $x \neq x'$. But f has no fixed point.

Corollary 2.4.1. *Let $f : (X, d) \to (X, d)$ be a function on a complete
metric space (X, d). If f^n is a contraction for some $n \in \mathbb{N}$, then f
has a unique fixed point in X.*

Proof. By Theorem 2.4.1, f^n has a unique fixed point say x_o. Thus,
$f^n(x_o) = x_o$. Now

$$f^n(f(x_o)) = f(f^n(x_o)) = f(x_o).$$

This implies that $f(x_o)$ is a fixed point of f^n. Then by the uniqueness
of the fixed point of f^n, $f(x_o) = x_o$. Thus, x_o is a fixed point of f.
Moreover, it is unique: if $f(x_1) = x_1$ then $f^n(x_1) = x_1$. Since x_o is
the unique fixed point of f^n, $x_1 = x_o$. $\square$

We end this section with an interesting observation related to
continuous maps between normed linear spaces. Note that a metric
can be defined on any non-empty set X. But if the set X possesses
a linear structure (that is, X is a vector space) then we can think
of combining the two structures such that the two operations: vector
addition and scalar multiplication on X are continuous. Such a vector
space, endowed with a metric, in which vector addition and scalar

multiplication are continuous is known as a **metric linear space**. In fact, we usually consider normed linear spaces which are special cases of metric linear spaces (Exercise 2.4.13).

Since a normed linear space is also a vector space, we can also think about the functions which preserve the linear structure of the normed linear spaces. Such maps play a crucial role in functional analysis.

Definition 2.4.4. Let X and Y be vector spaces over $\mathbb{K}$, where $\mathbb{K} = \mathbb{R}$ or $\mathbb{C}$. Then a function $T : X \to Y$ is called a **linear transformation** (or a linear operator) if for all x_1, $x_2 \in X$ and for all α, $\beta \in \mathbb{K}$,

$$T(\alpha x_1 + \beta x_2) = \alpha T(x_1) + \beta T(x_2).$$

Note. *If T is a linear transformation, then $T(0) = 0$. (For convenience, we denote the zero vector in X and that in Y by the same notation.)*

When we are dealing with a linear transformation between two normed linear spaces, then we can also talk about its continuity. Interestingly, if a linear transformation is continuous at a single point then it is continuous on the whole domain space. Moreover, continuity and linearity of a function together produces the Lipschitz property. Let us precisely state this in the next theorem.

Theorem 2.4.2. *Let $T : (X, \|\cdot\|) \to (Y, \|\cdot\|)$ be a linear transformation between two normed linear spaces. Then the following statements are equivalent:*

(a) *T is uniformly continuous.*
(b) *T is continuous.*
(c) *T is continuous at a point $x_o \in X$.*
(d) *T is continuous at 0.*
(e) *There exists $M > 0$ such that $\|Tx\| \leq M\|x\| \; \forall \; x \in X$.*
(f) *There exists $M' > 0$ such that $\|Tx - Tx'\| \leq M'\|x - x'\| \; \forall \; x, x' \in X$.*

Remarks. (1) For convenience, we are denoting the norms on X and Y by the same symbol $\|\cdot\|$.
(2) Condition (f) implies that T is Lipschitz.
(3) Continuous linear maps between normed linear spaces are generally referred to as *bounded linear operators*.

Proof. The implications (a) $\Rightarrow$ (b) $\Rightarrow$ (c) are immediate.

(c) $\Rightarrow$ (d): We use Theorem 2.1.2 to prove the assertion. Let $x_n \to$ 0 in X. Therefore, $(x_n + x_o) \to x_o$ and consequently, $T(x_n + x_o) \to T(x_o)$. Since T is linear, $Tx_n \to 0$ in Y.

(d) $\Rightarrow$ (e): Let $\epsilon > 0$. Since T is continuous at 0, there exists a $\delta > 0$ such that $\|x\| < \delta$ implies $\|Tx\| < \epsilon$. Let $0 \neq x_o \in X$. Now $\left\|\frac{\delta}{2}\frac{x_o}{\|x_o\|}\right\| < \delta$, therefore $\left\|T\left(\frac{\delta}{2}\frac{x_o}{\|x_o\|}\right)\right\| < \epsilon$. Thus we have $\|T(x_o)\| < \frac{2\epsilon}{\delta}\|x_o\|$. Let $M = \frac{2\epsilon}{\delta} (> 0)$, we get $\|Tx\| \leq M\|x\| \; \forall \; x \in X$.

(e) $\Rightarrow$ (f): Let $x, x' \in X$. From (e), we have $\|T(x-x')\| \leq M\|(x-x')\|$. Thus taking $M' = M$, we get $\|Tx - Tx'\| \leq M'\|x - x'\| \; \forall \; x, x' \in X$.

(f) $\Rightarrow$ (a): This follows immediately from the fact that Lipschitz function is uniformly continuous. $\qquad\square$

Corollary 2.4.2. *Let $T : (X, \|\cdot\|) \to (Y, \|\cdot\|)$ be a linear transformation between two normed linear spaces. Suppose x, $x' \in X$. Then T is continuous at x if and only if T is continuous at x'.*

Exercises

2.4.1. Let $(X, \|\cdot\|)$ be a normed linear space. Show that the norm function $f : X \to \mathbb{R} : f(x) = \|x\|$ is Lipschitz.

2.4.2. Suppose (X, d) is a metric space. Prove that the distance function $f : (X \times X, \mu_1) \to \mathbb{R} : f(x, y) = d(x, y)$ is Lipschitz. Is it true if we replace μ_1 with μ_2 or μ_∞ (refer to Example 1.3.4)?

2.4.3. Show that every polynomial function is Lipschitz on any bounded interval in $\mathbb{R}$.

2.4.4. Let (X, d) be a metric space and $F = \{f : (X, d) \to \mathbb{R} : f \text{ is } M\text{-Lipschitz}\}$ for some $M > 0$. For every $x \in X$, let $\phi(x) = \inf_{f \in F} f(x) \in \mathbb{R}$. Prove that the function ϕ is M-Lipschitz. Similarly, prove that the function $\psi(x) = \sup_{f \in F} f(x)$ is M-Lipschitz provided ψ is real-valued for all $x \in X$.

2.4.5. Show that Lipschitz functions preserve boundedness, that is, if $f : (X, d) \to (Y, \rho)$ is a Lipschitz function and $A \subseteq X$ is bounded, then $f(A)$ is bounded in (Y, ρ). Using this observation, can you construct an example of a non-Lipschitz function?

2.4.6. Let (X, d) be a metric space. If the functions f, $g : (X, d) \to \mathbb{R}$ are bounded and Lipschitz, then prove that their product fg is also Lipschitz. In fact, using previous exercise, one can say that the product of two real-valued Lipschitz functions defined on a bounded metric space is also Lipschitz. Think of an example of two Lipschitz functions whose product is not Lipschitz.

2.4.7. If $f : I \to \mathbb{R}$ is continuous on an interval I of $\mathbb{R}$ and differentiable in $\text{int}(I)$, then f' is bounded on $\text{int}(I)$ if and only if f is Lipschitz.

2.4.8. Let $\{(X_i, d_i) : 1 \le i \le n\}$ be a finite collection of metric spaces and $\pi_i : (\prod_{j=1}^{n} X_j, \mu_\infty) \to (X_i, d_i)$ be the projection map of $\prod_{j=1}^{n} X_j$ onto X_i for $i \in \{1, 2, \ldots, n\}$. Show that π_i is Lipschitz. Observe that π_i is Lipschitz even if μ_∞ is replaced by μ_1 or μ_2 (refer to Example 1.3.4).

2.4.9. Let A be a non-empty closed subset of $\mathbb{R}$. If $x \in \mathbb{R}$, then prove that $d(x, A) = d(x, a)$ for some $a \in A$.

2.4.10. Give an example of a uniformly continuous function which is not Lipschitz.

2.4.11. Suppose $\text{Lip}(X)$ denotes the collection of all real-valued Lipschitz functions on a metric space (X, d). It is easy to verify that $\text{Lip}(X)$ forms a vector space with respect to the usual addition of functions on X and the usual scalar multiplication. Let $x_o \in X$. Show that $\| \cdot \|$ defines a norm on $\text{Lip}(X)$, where $\| \cdot \|$ is given by

$$\|f\| = \max\{|f(x_o)|, \ L(f)\} \ \text{ for } f \in \text{Lip}(X).$$

2.4.12. A function $f : (X, d) \to (Y, \rho)$ is said to be *Hölder continuous* if there exist $M > 0$ and $\alpha > 0$ such that

$$\rho(f(x), f(x')) \le M \ (d(x, x'))^\alpha \text{ for all } x, \ x' \in X.$$

Clearly, the class of Lipschitz functions is contained in the class of Hölder continuous functions with $\alpha = 1$. Establish the following assertions:

(a) Every Hölder continuous function is uniformly continuous.
(b) There exists an Hölder continuous function which is not Lipschitz.
(c) Suppose I is any interval in $\mathbb{R}$ and $f : I \to \mathbb{R}$ is Hölder continuous with $\alpha > 1$. Then f is constant on I.

2.4.13. Let $(X, \|\cdot\|)$ be a normed linear space over $\mathbb{K}$, where $\mathbb{K} = \mathbb{R}$ or $\mathbb{C}$. Show that the vector addition and scalar multiplication on X are continuous operations, that is, the functions $f_1 : (X \times X, \mu_\infty) \to (X, d)$: $f_1((x, y)) = x + y$ and $f_2 : (\mathbb{K} \times X, \mu_\infty) \to (X, d)$: $f_2((\lambda, x)) = \lambda \cdot x$ are continuous. Here observe that d denotes the metric induced by the norm $\|\cdot\|$, that is, $d(x, y) = \|x - y\|$. Further, note that the functions f_1 and f_2 are continuous even if μ_∞ is replaced by μ_1 or μ_2 (refer to Example 1.3.4).

2.5   Isometries between Metric Spaces

Recall that in the beginning of this chapter, we were talking about the vehicle which could carry some properties from one metric space to another. Now let us discuss regarding a very strong version of continuity. Whatever forms of continuity we have discussed so far does not preserve the distance, that is, the distance between the elements of the domain space need not be same as the distance between the respective images. But since we are dealing with the metric structures, we should also look into this possibility so that the corresponding range space $(f(X), \rho)$ is just a copy of the domain space (X, d) which is obtained by renaming the elements of X.

Before starting with the definition of an isometry between metric spaces, let us mention the following simple result which we will need later.

Proposition 2.5.1. *Let (X, d) be a metric space and let (x_n) and (y_n) be two sequences in X. Then*

(a) $|d(x_n, y_n) - d(x_m, y_m)| \leq d(x_n, x_m) + d(y_n, y_m)$
(b) *If (x_n) and (y_n) are Cauchy in (X, d), then $(d(x_n, y_n))$ is a Cauchy sequence in $(\mathbb{R}, |\cdot|)$ and consequently it converges in $(\mathbb{R}, |\cdot|)$.*

(c) If $x_n \to x$, $y_n \to y$ in (X, d), then $d(x_n, y_n) \to d(x, y)$ in $(\mathbb{R}, |\cdot|)$. In notation, $\lim\limits_{n \to \infty} d(x_n, y_n) = d(x, y) = d(\lim\limits_{n \to \infty} x_n, \lim\limits_{n \to \infty} y_n)$.
(Note the interchange of lim and d.)

Definition 2.5.1. A function $T : (X, d) \to (Y, \rho)$ between two metric spaces is called an **isometry** if $\rho(Tx, Tx') = d(x, x')$ holds for all x, $x' \in X$.

Note that an isometry is always one-to-one: $Tx = Tx' \Rightarrow \rho(Tx, Tx') = 0 \Rightarrow d(x, x') = 0 \Rightarrow x = x'$. But an isometry need not be onto: consider the map $f : (\mathbb{N}, |\cdot|) \to (\mathbb{N}, |\cdot|)$ defined by $f(n) = n + 1$. Since 1 does not have a pre-image, f is not onto. If an onto isometry exists between (X, d) and (Y, ρ), then the two metric spaces are called **isometric**. Evidently, isometric metric spaces are homeomorphic. Moreover, every isometry is Lipschitz and hence uniformly continuous as well.

Note. *If there exists an onto isometry from (X, d) to (Y, ρ) then the inverse map is an onto isometry from (Y, ρ) to (X, d). Hence we just call the metric spaces to be isometric (without mentioning from where to where).*

Theorem 2.5.1. *Let (X, d) and (Y, ρ) be two complete metric spaces such that X' is dense in X, while Y' is dense in Y. Suppose $T' : (X', d) \to (Y', \rho)$ is an onto isometry. Then there exists an onto isometry $T : (X, d) \to (Y, \rho)$ such that T is an extension of T', that is, $Tx = T'x \ \forall \ x \in X'$.*

Proof. Let $x \in X$. Since X' is dense in X, $\exists$ a sequence (x_n) in X' such that $x_n \to x$. Now $d(x_n, x_m) = \rho(T'x_n, T'x_m) \ \forall \ n, \ m \in \mathbb{N}$. But (x_n) is Cauchy and hence $(T'x_n)$ is also Cauchy in (Y, ρ). Since (Y, ρ) is complete, $T'x_n \to y$ for some $y \in Y$. We can define $Tx = y = \lim\limits_{n \to \infty} T'x_n$.

T *is well-defined:* We need to ensure that this definition does not depend on the choice of the sequence (x_n) converging to x. So let (z_n) be another sequence in X converging to x in (X, d). Now by Proposition 2.5.1,

$$\rho(\lim\limits_{n \to \infty} T'x_n, \lim\limits_{n \to \infty} T'z_n) = \lim\limits_{n \to \infty} \rho(T'x_n, T'z_n)$$

$$= \lim\limits_{n \to \infty} d(x_n, z_n)$$

$$= d(\lim_{n\to\infty} x_n, \lim_{n\to\infty} z_n)$$

$$= d(x, x) = 0.$$

Hence $\lim_{n\to\infty} T'x_n = \lim_{n\to\infty} T'z_n$. It follows that $Tx = \lim_{n\to\infty} T'x_n$ is a well-defined function from X to Y.

Clearly, $Tx = T'x$ for $x \in X$ (consider the constant sequence). Now we show that $T : X \to Y$ is an isometry. Let $x, z \in X$. Suppose that (x_n) and (z_n) are two sequences in (X', d) converging to x and z respectively. Again by Proposition 2.5.1, we have

$$\rho(Tx, Tz) = \rho(\lim_{n\to\infty} T'x_n, \lim_{n\to\infty} T'z_n)$$

$$= \lim_{n\to\infty} \rho(T'x_n, T'z_n)$$

$$= \lim_{n\to\infty} d(x_n, z_n)$$

$$= d(\lim_{n\to\infty} x_n, \lim_{n\to\infty} z_n) = d(x, z).$$

Finally, we need to show that T is onto. Let $y \in Y$. Since Y' is dense in Y, $\exists$ a sequence (y_n) in Y' such that $y_n \to y$ in (Y, ρ). So (y_n) is Cauchy in (Y', ρ). Since $T' : (X', d) \to (Y', \rho)$ is an onto isometry, $(T')^{-1} : (Y', \rho) \to (X', d)$ is also an isometry. Hence $\rho(y_n, y_m) = d((T')^{-1}y_n, (T')^{-1}y_m)$ and consequently the sequence $\{(T')^{-1}(y_n)\}$ is Cauchy in (X, d). Let $(T')^{-1}(y_n) = z_n$. Then $T'z_n = y_n$. Since (X, d) is complete, $z_n \to x$ for some $x \in X$. Thus, $Tz_n \to Tx$. But $Tz_n = T'z_n \ \forall \ n$, since T is an extension of T'. So $Tz_n = T'z_n = y_n \to y \Rightarrow y = Tx$. Hence T is onto. $\qquad\square$

Remark. It is easy to see that under the hypothesis of the previous theorem, there exists a unique isometry from X onto Y which extends T': suppose $T_1 : (X, d) \to (Y, \rho)$ is another onto isometry such that $T_1 x = T'x \ \forall \ x \in X'$. Let $x \in X$. Then there exists $(x_n) \subseteq X'$ such that $x_n \to x$. Then by the continuity of T_1, $T_1(x_n) \to T_1(x)$. Therefore, $T_1(x) = \lim_{n\to\infty} T_1(x_n) = \lim_{n\to\infty} T'(x_n) = T(x)$.

Exercises

2.5.1. Prove Proposition 2.5.1.

2.5.2. Prove that the interval $[0, 1]$ is isometric to any closed interval of length 1. Is it isometric to any closed interval of length other than 1?

2.5.3. Suppose $\mathcal{A}$ is a family of metric spaces. Let us define a relation $\sim$ on $\mathcal{A}$ as follows: for (X, d) and (Y, ρ) in $\mathcal{A}$,

$$X \sim Y \iff X \text{ and } Y \text{ are isometric.}$$

Show that $\sim$ is an equivalence relation, that is, it is reflexive, symmetric and transitive.

2.5.4. (a) Let $f : (X, d) \to (Y, \rho)$ be an onto isometry. If $x_o \in X$, $r > 0$, $A = \{x \in X : d(x, x_o) = r\}$ and $B = \{y \in Y : \rho(y, f(x_o)) = r\}$, then show that $B = f(A)$.
(b) Use (a) to show that there does not exist any onto isometry between $(\mathbb{R}, |\cdot|)$ and $(\mathbb{R}^2, |\cdot|)$.

2.5.5. Prove that every separable metric space is isometric to a subset of ℓ^∞.

2.5.6. Produce an isometry of $\mathbb{R}^n$ equipped with the Euclidean metric into ℓ^2.

2.5.7. For a metric space (X, d), show that $x \mapsto \{x\}$ is an isometry of X into the collection of all non-empty closed and bounded subsets of X equipped with the Hausdorff distance. (Refer to Exercise 1.5.7.)

2.6 Equivalent Metrics

Many a times working with a particular metric could be little cumbersome than other metrics. In case we are looking for certain topological features which only involves the open sets, etc., then we can work with some metric which is *equivalent* to the original metric. Before we give the precise meaning of equivalent metrics, let us start with the following relevant result. Its proof is left as an exercise.

Theorem 2.6.1. *Let d and ρ be two metrics on a non-empty set X. Then the following statements are equivalent:*

(a) *The family τ_d of all open sets in X generated by d is same as the family τ_ρ of all open sets generated by ρ.*

(b) *Both the identity map $I : (X, d) \to (X, \rho)$ and its inverse I^{-1} are continuous. In other words, the identity map is a homeomorphism.*

(c) *If (x_n) is a sequence in X and $x \in X$ then $\lim\limits_{n\to\infty} d(x_n, x) = 0$ if and only if $\lim\limits_{n\to\infty} \rho(x_n, x) = 0$, that is, a sequence (x_n) in X converges to some x in (X, d) if and only if (x_n) converges to x in (X, ρ).*

Definition 2.6.1. When any one of the equivalent conditions of the previous theorem is satisfied, then the metrics d and ρ are called **equivalent metrics**.

Example 2.6.1. If d is a metric on X, then the metric ρ defined by

$$\rho(x, y) = \frac{d(x, y)}{1 + d(x, y)}, \quad x, \ y \in X$$

is equivalent to d.

Let $\lim\limits_{n\to\infty} d(x_n, x) = 0$. Then it is clear that $\rho(x_n, x) \to 0$ as $n \to \infty$. Conversely, let $\lim\limits_{n\to\infty} \rho(x_n, x) = 0 \ -(\star)$. Since $\rho(x_1, x_2) < 1 \ \forall \ x_1, x_2 \in X$, $d(x_n, x) = \frac{\rho(x_n, x)}{1 - \rho(x_n, x)} \to 0$ as $n \to \infty$ by $(\star)$.

Example 2.6.2. On $X = (-\frac{\pi}{2}, \frac{\pi}{2})$, consider a metric ρ given by

$$\rho(x, y) = |\tan x - \tan y|.$$

Then ρ is a metric on $(-\frac{\pi}{2}, \frac{\pi}{2})$ which is equivalent to the usual distance metric $| \cdot |$ on $(-\frac{\pi}{2}, \frac{\pi}{2})$ by the continuity of $\tan x$, $\tan^{-1} x$ and Theorem 2.6.1(c). Also see Exercise 2.6.4. Note that $(-\frac{\pi}{2}, \frac{\pi}{2})$ is incomplete with respect to the usual metric $|\cdot|$ because the Cauchy sequence $(\frac{\pi}{2} - \frac{1}{n})$ does not converge in $(X, |\cdot|)$. But (X, ρ) is complete: let (x_n) be a Cauchy sequence in (X, ρ). This implies that $(\tan(x_n))$ is Cauchy in $(\mathbb{R}, |\cdot|)$. Since $(\mathbb{R}, |\cdot|)$ is complete, the sequence $(\tan(x_n))$ converges to some $a = \tan(\tan^{-1} a)$ in $(\mathbb{R}, |\cdot|)$. Consequently, $\rho(x_n, \tan^{-1} a) \to 0$ as $n \to \infty$. Hence (x_n) converges to $\tan^{-1} a$ in (X, ρ).

Remark. Note that in the previous example, $\tan x$ function has been used because it defines a homeomorphism from the incomplete metric space $((-\frac{\pi}{2}, \frac{\pi}{2}), |\cdot|)$ to the complete metric space $(\mathbb{R}, |\cdot|)$.

In view of the previous example, we would like to make the following observation:

Two metrics d and ρ on a non-empty set X may be equivalent, but one of them may be complete, while the other one may be incomplete. The equivalence of two metrics only guarantees the same family of convergent sequences with same limit, but the equivalent metrics need not generate the same family of Cauchy sequences. It should be noted that convergence of a sequence depends on the open sets, while Cauchyness rely on the size of the balls.

Example 2.6.3. Consider two metrics d_1 and d_2 on the set $(0, \infty)$, where d_1 is the usual distance metric and d_2 is defined as follows:

$$d_2(x, y) = \left| \frac{1}{x} - \frac{1}{y} \right| \quad \text{for } x, \ y \in (0, \infty).$$

It is easy to see that d_1 and d_2 are equivalent metrics (see Exercise 2.6.4) and hence they have the same collection of convergent sequences. But the sequence $(\frac{1}{n})$ is Cauchy with respect to d_1 but not with respect to d_2.

In order to get rid of this anomaly, we need a stronger concept than that of equivalent metrics.

Definition 2.6.2. Let d and ρ be two metrics on a non-empty set X. Suppose both the identity map $I : (X, d) \to (X, \rho)$ and its inverse I^{-1} are uniformly continuous. Then d and ρ are called **uniformly equivalent**.

Example 2.6.4. If $f : (X, d) \to (Y, \mu)$ is uniformly continuous, then the metric ρ on X defined as: $\rho(x_1, x_2) = d(x_1, x_2) + \mu(f(x_1), f(x_2))$ for all $x_1, x_2 \in X$, is uniformly equivalent to d.

Example 2.6.5. Refer to Example 1.3.4 for metrics defined on the Cartesian product $\prod_{i=1}^{n} X_i$ of the finite collection of metric spaces $\{(X_i, d_i) : 1 \leq i \leq n\}$. Recall that

$$\mu_\infty(a, b) \leq \mu_2(a, b) \leq \mu_1(a, b) \leq n\mu_\infty(a, b),$$

for $a = (a_1, \ldots, a_n)$, $b = (b_1, \ldots, b_n) \in \prod_{i=1}^{n} X_i$. Thus, both the metrics μ_1 and μ_2 are uniformly equivalent (and hence equivalent) to μ_∞. Consequently, whenever we need to verify some property, possessed by the Cartesian product, which involves only the open sets then it is enough to work with any one of these metrics at our convenience. The topology generated by these metrics is called the *product topology* and hence the Cartesian product equipped with the product topology is referred to as the *product space*. (See Exercise 2.6.7.)

Proposition 2.6.1. *Given any metric space* (X, d), *there exists a metric* ρ *on* X *such that* ρ *is uniformly equivalent to* d *and* (X, ρ) *is bounded.*

Proof. We prove that the metric ρ given in Example 2.6.1 is in fact uniformly equivalent to d. First observe that $\rho(x, y) = \frac{d(x,y)}{1+d(x,y)} \leq d(x, y)$ for all x, $y \in X$. Then, the identity map from (X, d) to (X, ρ) is Lipschitz and hence uniformly continuous. Now let $\epsilon > 0$. Choose $\delta = \frac{\epsilon}{1+\epsilon}$. Then

$$\rho(x, y) < \delta \Rightarrow 1 - \frac{1}{1 + d(x, y)} < 1 - \frac{1}{1 + \epsilon} \Rightarrow d(x, y) < \epsilon.$$

Thus, the identity map from (X, ρ) to (X, d) is uniformly continuous. Now it is proved that ρ is uniformly equivalent to d. Further, it is evident that $\rho(x, y) \leq 1$ for all x, $y \in X$. Hence (X, ρ) is bounded. $\square$

As a consequence of the previous result, it is easy to deduce that boundedness is not preserved under uniform equivalence. Since the identity maps between the uniformly equivalent metrics are uniformly continuous, both the spaces have the same class of Cauchy sequences and that of convergent sequences. Now the next result immediately follows from this observation.

Proposition 2.6.2. *Let* d *and* ρ *be two uniformly equivalent metrics on a non-empty set* X. *Then* (X, d) *is complete if and only if* (X, ρ) *is complete.*

Let us end this section with interesting remarks. Carefully observe the definition of uniformly equivalent metrics. *Can you define a notion which is stronger than homeomorphism?*

Definition 2.6.3. A function $f : (X, d) \to (Y, \rho)$ between two metric spaces is said to be a **uniform homeomorphism** between (X, d) and (Y, ρ) if f is a bijection such that both f and f^{-1} are uniformly continuous. Moreover, if such a uniform homeomorphism exists then the metric spaces (X, d) and (Y, ρ) are called **uniformly homeomorphic**.

Remarks. (a) Unlike in homeomorphism, completeness is preserved under uniform homeomorphism, that is, if (X, d) and (Y, ρ) are uniformly homeomorphic and if (X, d) is complete then (Y, ρ) is also complete. This follows immediately from Proposition 2.3.1.

(b) Two metrics d and ρ on a non-empty set X are uniformly equivalent if and only if the identity map $I : (X, d) \to (X, \rho)$ is a uniform homeomorphism.

(c) By Proposition 2.6.1 and the previous remark, it can be deduced that boundedness may not be preserved under uniform homeomorphism.

Exercises

2.6.1. Prove Theorem 2.6.1.

2.6.2. Let $\{f_1, \ldots, f_n\}$ be a finite collection of functions where $f_i : (X, d) \to (Y_i, \rho_i)$ for each i. Suppose d' is a metric on X which is defined as

$$d'(x_1, x_2) = d(x_1, x_2) + \sum_{i=1}^{n} \rho_i(f_i(x_1), f_i(x_2)) \text{ for all } x_1, x_2 \in X.$$

Show that d' is equivalent to d if and only if $f_i : (X, d) \to (Y_i, \rho_i)$ is continuous for each $i \in \{1, 2, \ldots, n\}$. Moreover, in this case $f_i : (X, d') \to (Y_i, \rho_i)$ is Lipschitz for each i.

2.6.3. Let us call the metrics d and ρ on a non-empty set X to be **Cauchy equivalent** if both the identity map $I : (X, d) \to (X, \rho)$ and its inverse I^{-1} are Cauchy-continuous. Hence, the collection of Cauchy sequences in (X, d) coincides with that of (X, ρ). Give an example of Cauchy-equivalent metrics which are not uniformly equivalent.

2.6.4. Let $f : (X, d_1) \to (Y, \rho)$ be a bijective function. Define a metric d_2 on X as follows: $d_2(x, x') = \rho(f(x), f(x'))$. Then show that the metrics d_1 and d_2 are equivalent on X if and only if f is a homeomorphism. Can you think of some equivalent conditions for Cauchy-equivalence and uniform equivalence of metrics?

2.6.5. Produce a metric ρ on $X = (-1, 1)$ such that ρ is equivalent to the usual metric and (X, ρ) is complete. (Also see Theorem 3.4.3.)

2.6.6. Show that the metric ρ on X, defined as $\rho(x_1, x_2) = \min\{1, d(x_1, x_2)\}$ for $x_1,\ x_2 \in X$, is uniformly equivalent to the metric d on X.

2.6.7. Prove that the open sets in the product topology on $\prod_{i=1}^{n} X_i$ are precisely the arbitrary unions of the sets of the form $\prod_{i=1}^{n} U_i$, where U_i is open in (X_i, d_i) for $i \in \{1, 2, \ldots, n\}$.

2.6.8. Let $C[0, 1]$ be the set of all real-valued continuous functions defined on the interval $[0, 1]$. Show that the metrics d and ρ on $C[0, 1]$, defined as

$$d(f, g) = \max\{|f(x) - g(x)| : x \in [0, 1]\} \text{ and}$$

$$\rho(f, g) = \int_0^1 |f(x) - g(x)| dx,$$

for $f,\ g \in C[0, 1]$, are not equivalent.

2.6.9. Prove that $\mathbb{R}$ is not uniformly homeomorphic with $(0, 1)$. Compare with Example 2.3.1.

Chapter 3

Complete Metric Spaces

If a metric space possesses some special properties, then such metric spaces become more useful. One of such significant properties is completeness which we are going to study in this chapter. Due to various applications of complete metric spaces, one usually look for the completion of an incomplete metric space. It is, in some sense, the 'smallest' complete metric space in which an incomplete metric space is contained. The interesting process is discussed in detail in this chapter.

3.1 Some Examples

Consider the metric space $(\mathbb{Q}, |\cdot|)$, where $\mathbb{Q}$ is the set of all rational numbers and $|\cdot|$ is the usual distance on $\mathbb{Q}$. Define a sequence (x_n) in $\mathbb{Q}$ as follows: let $x_1 = 1$ and then define inductively $x_{n+1} = \frac{1}{2}(x_n + \frac{2}{x_n})$ for all n. By induction, we have (x_n) to be a sequence of rationals. But where does (x_n) converge to? The sequence (x_n) converges to $\sqrt{2}$ (Prove it!). But $\sqrt{2}$ is not a rational number. This means (x_n) does not converge in $(\mathbb{Q}, |\cdot|)$. Since we know that the members of (x_n) actually gather around $\sqrt{2}$ after a while, the members of (x_n) get close to each other eventually. This much information we have about (x_n) as long as we confine ourselves to the smaller metric space $(\mathbb{Q}, |\cdot|)$ instead of the bigger space $(\mathbb{R}, |\cdot|)$. But for a general metric space (X, d), we may not know if it is sufficiently large or not. This leads to the definition of a special class of sequences which is weaker than that of convergent sequences.

Definition 3.1.1. A sequence (x_n) in a metric space (X, d) is called a Cauchy sequence if for every $\epsilon > 0$, there exists $n_\epsilon \in \mathbb{N}$ (n_ϵ depends on ϵ) such that $d(x_m, x_n) < \epsilon$ for all n, $m \geq n_\epsilon$.

Thus every convergent sequence is Cauchy. But the discussion before the previous definition depicts that the sequence (x_n) was Cauchy but not convergent in $(\mathbb{Q}, |\cdot|)$. There we observed a crucial property which is possessed by $(\mathbb{R}, |\cdot|)$ but not by $(\mathbb{Q}, |\cdot|)$.

Definition 3.1.2. If every Cauchy sequence in (X, d) converges in (X, d), then (X, d) is called a complete metric space.

We will shortly prove that $(\mathbb{R}, |\cdot|)$ is a complete metric space. Before that let us look at some examples of complete and incomplete metric spaces.

Example 3.1.1. Let $X = (0, \infty)$ be equipped with the usual metric $d(x, y) = |x - y|$. Then (X, d) is not complete as $\left(\frac{1}{n}\right)$ is a Cauchy sequence in (X, d) which does not converge in (X, d) ($0 \notin X$).

Example 3.1.2. Let (X, d) be a metric space where d is the discrete metric. Then every Cauchy sequence in (X, d) is eventually constant and hence it converges. Thus (X, d) is a complete metric space. What can you say about the metric space $(\mathbb{N}, |\cdot|)$?

Before we discuss about a few more important examples of complete metric spaces, let us briefly highlight one result from real analysis which could be useful in proving the completeness of certain metric spaces.

Proposition 3.1.1. *Let (x_n) be a sequence in $\mathbb{R}$ converging to a point x in $\mathbb{R}$. Then the sequence $(|x_n|)$ converges to $|x|$ in $\mathbb{R}$.*

Proof. Use $||x_n| - |x|| \leq |x_n - x|$. $\qquad\square$

Is the converse of the previous result true? That is, if $|x_n| \to |x|$ in $(\mathbb{R}, |\cdot|)$, then does that imply $x_n \to x$?

Example 3.1.3. Consider l^p ($p \geq 1$ is a fixed real number) equipped with the metric

$$d(x, y) = \left(\sum_{k=1}^{\infty} |x_k - y_k|^p \right)^{1/p},$$

where $x = (x_k)$ and $y = (y_k)$ (refer to Example 1.3.7). We now prove that it is a complete metric space.

Let $(x^{(n)})$ be a Cauchy sequence in (l^p, d), where $x^{(n)} = (x_1^{(n)}, x_2^{(n)}, \ldots)$. Then for every $\epsilon > 0$, there exists $n_o \in \mathbb{N}$ such that for all $m, n > n_o$,

$$d(x^{(m)}, x^{(n)}) = \left(\sum_{j=1}^{\infty} |x_j^{(m)} - x_j^{(n)}|^p \right)^{1/p} < \epsilon. \quad (\star)$$

Hence for every j, $|x_j^{(m)} - x_j^{(n)}| < \epsilon$, for all $m, n > n_o$. This implies that the sequence $(x_j^{(n)})_{n \in \mathbb{N}}$ is a Cauchy sequence in $\mathbb{F}$, where $\mathbb{F} = \mathbb{R}$ or $\mathbb{C}$, for every j and therefore it converges to some x_j, because $\mathbb{F}$ with the usual metric is complete.

Let $x = (x_1, x_2, \ldots)$. Now we claim that $x \in l^p$ and the sequence $(x^{(n)})$ converges to x. From $(\star)$, for $k \in \mathbb{N}$ we have

$$\sum_{j=1}^{k} |x_j^{(m)} - x_j^{(n)}|^p < \epsilon^p$$

for all $m, n > n_o$. If $n \to \infty$, then

$$\sum_{j=1}^{k} |x_j^{(m)} - x_j|^p \le \epsilon^p \quad \text{for} \quad m > n_o.$$

Therefore, the sequence of partial sums (s_k), where $s_k = \sum_{j=1}^{k} |x_j^{(m)} - x_j|^p$, is monotonically increasing and bounded above and hence we get,

$$\sum_{j=1}^{\infty} |x_j^{(m)} - x_j|^p \le \epsilon^p \quad \text{for} \quad m > n_o. \quad (\star\star)$$

Consequently, $x^{(m)} - x \in l^p$. Since $x^{(m)} \in l^p$, by Minkowski's inequality (Theorem 7.4.6), we have

$$x = x^{(m)} + (x - x^{(m)}) \in l^p.$$

Moreover, by $(\star\star)$ we have $(x^{(n)})$ converges to x.

Similar steps can be used to prove the completeness of the following metric space.

Example 3.1.4. Consider the space l^∞ equipped with the supremum metric $d(x,y) = \sup_{n \in \mathbb{N}} |x_n - y_n|$ (refer to Example 1.3.8). Then the metric space (l^∞, d) is complete.

It should be noted that the metric plays a crucial role in the completeness of a metric space.

Example 3.1.5. Consider the space $C[a,b]$ of all continuous real-valued functions on a closed and bounded interval $[a,b]$ equipped with the uniform metric:

$$D(x,y) = \max_{t \in [a,b]} |x(t) - y(t)|,$$

where x, y are continuous functions on $[a,b]$. Then $C[a,b]$ is complete (proved later). But if we equip the set with the metric:

$$d_p(x,y) = \left(\int_a^b |x(t) - y(t)|^p dt \right)^{1/p},$$

where $1 \leq p < \infty$ then the metric space is incomplete. We now show the incompleteness of the space $(C[a,b], d_1)$. First observe that $d_1(x,y)$ geometrically represents the area between the graphs of the functions x and y. Now let $a < c < b$ and for every n with $a < c - 1/n$, define x_n as

$$x_n(t) = \begin{cases} 0 & \text{if } a \leq t \leq c - \frac{1}{n} \\ nt - nc + 1 & \text{if } c - \frac{1}{n} \leq t \leq c \\ 1 & \text{if } c \leq t \leq b \end{cases}.$$

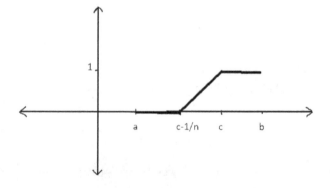

Then (x_n) is a Cauchy sequence in $(C[a,b], d_1)$ as $d_1(x_n, x_m) \leq \frac{1}{n} + \frac{1}{m}$. Suppose (x_n) converges to some $x \in C[a,b]$. Now

$$d_1(x_n, x) = \int_a^{c-(1/n)} |x(t)|dt + \int_{c-(1/n)}^c |x_n(t) - x(t)|dt$$

$$+ \int_c^b |1 - x(t)|dt$$

$$= \left(\int_a^{c-(1/n)} |x(t)|dt + \int_{c-(1/n)}^c |x(t)|dt \right)$$

$$+ \int_{c-(1/n)}^c |x_n(t) - x(t)|dt + \int_c^b |1 - x(t)|dt$$

$$- \int_{c-(1/n)}^c |x(t)|dt$$

$$= \int_a^c |x(t)|dt + \int_{c-(1/n)}^c |x_n(t) - x(t)|dt$$

$$+ \int_c^b |1 - x(t)|dt - \int_{c-(1/n)}^c |x(t)|dt.$$

Since $c - \frac{1}{n} \to c$, the integrals $\int_{c-(1/n)}^c |x_n(t) - x(t)|dt$ and $\int_{c-(1/n)}^c |x(t)|dt$ converge to zero. Moreover, since $d_1(x_n, x) \to 0$, we have

$$\int_a^c |x(t)|dt + \int_c^b |1 - x(t)|dt = 0.$$

But $x \in C[a,b]$. This implies that $x(t) = 0$ on $[a,c]$ and $x(t) = 1$ on $[c,b]$, which is impossible. Hence the Cauchy sequence (x_n) does not converge in $(C[a,b], d_1)$.

Example 3.1.6. Let X be a non-empty set and let $B(X)$ be the set of all real-valued bounded functions defined on X equipped with the supremum metric D:

$$D(f,g) = \sup\{|f(x) - g(x)| : x \in X\}.$$

Now we prove that D is a complete metric: let (f_n) be a Cauchy sequence in $(B(X), D)$. Then given $\epsilon > 0$, there exists $n_\epsilon \in \mathbb{N}$ such that $D(f_n, f_m) < \epsilon \ \forall \ n, \ m \geq n_\epsilon$. In particular, for each $x \in X$,

$$|f_n(x) - f_m(x)| \leq D(f_n, f_m) < \epsilon \ \forall \ n, \ m \geq n_\epsilon. \quad (\star)$$

But this means that for each $x \in X$, the sequence $(f_n(x))$ is Cauchy in $(\mathbb{R}, |\cdot|)$. Hence $(f_n(x))$ converges to a (unique) point, say y_x in $(\mathbb{R}, |\cdot|)$ (y_x depends on x). Define $f : X \to \mathbb{R}$ by: $f(x) = y_x = \lim\limits_{n\to\infty} f_n(x)$. Note that f is a well-defined real-valued function defined on X. By $(\star)$, we get

$$\lim_{n\to\infty} |f_n(x) - f_m(x)| \leq \epsilon \ \forall \ m \geq n_\epsilon. \quad (\star\star)$$

Now

$$\lim_{n\to\infty} |f_n(x) - f_m(x)| = |\lim_{n\to\infty} (f_n(x) - f_m(x))|$$

$$\text{(by Proposition 3.1.1)}$$

$$= |f(x) - f_m(x)|$$

$$\text{(by the definition of } f\text{)}.$$

So we get,

$$|f(x) - f_m(x)| \leq \epsilon \ \forall \ m \geq n_\epsilon \quad (\star\star\star)$$

Hence for all $x \in X$, we have $|f(x) - f_{n_\epsilon}(x)| \leq \epsilon$. Consequently, $|f(x)| \leq |f_{n_\epsilon}(x)| + \epsilon \ \forall \ x \in X$. Since $f_{n_\epsilon} \in B(X)$, there exists $M_\epsilon > 0$ such that $|f_{n_\epsilon}(x)| \leq M_\epsilon \ \forall \ x \in X$. Hence $|f(x)| \leq M_\epsilon + \epsilon = M$ (say) $\forall \ x \in X$, that is, $f \in B(X)$.
Now we need to show that $D(f, f_n) \to 0$ as $n \to \infty$. By $(\star\star\star)$, $|f(x) - f_m(x)| \leq \epsilon \ \forall \ m \geq n_\epsilon$ and $\forall \ x \in X$. So $D(f, f_m) \leq \epsilon \ \forall \ m \geq n_\epsilon$. But this precisely means that $f_n \to f$ in $(B(X), D)$. Hence $(B(X), D)$ is a complete metric space.

Remark. (a) In $(\star\star)$, $n \to \infty$ means that it is being done for sufficiently large n and in particular, for $n \geq n_\epsilon$; Moreover, in the limit process, '$<$' is replaced by '$\leq$' (for example, $\frac{1}{n} > 0 \ \forall \ n \in \mathbb{N}$, but $\lim\limits_{n\to\infty} \frac{1}{n} = 0$).

(b) Note that if we replace $\mathbb{R}$ by $\mathbb{C}$, that is, if we take $B(X)$ to be the set of all complex valued bounded functions on X, then analogously we can prove that $(B(X), D)$ is a complete metric space.

Surprisingly, if $B(X,Y)$ denotes the set of all bounded functions from a set X to a metric space (Y, ρ) and suppose (Y, ρ) is complete then $B(X,Y)$ is also complete with respect to the supremum metric. The proof is analogous to that of $B(X)$. Interestingly, the converse is also true, that is, if $(B(X,Y), D)$ is complete then so is (Y, ρ).

Proposition 3.1.2. *Let (X, d) and (Y, ρ) be two metric spaces such that $(B(X,Y), D)$ is complete, where $D(f, g) = \sup\{\rho(f(x), g(x)) : x \in X\}$. Then (Y, ρ) is also complete.*

Proof. Let (y_n) be a Cauchy sequence in Y. For each $n \in \mathbb{N}$, define $f_n : X \to Y$ as: $f_n(x) = y_n$. Then clearly, $f_n \in B(X,Y)$ for all $n \in \mathbb{N}$. Moreover, (f_n) is Cauchy in $B(X,Y)$. Since $(B(X,Y), D)$ is complete, let $f_n \to f$. Now it is enough to prove that f is a constant function. Suppose, if possible, $f(x_1) \neq f(x_2)$. Let $\epsilon_o = \rho(f(x_1), f(x_2)) > 0$. Since $f_n \to f$, there exists $n_o \in \mathbb{N}$ such that $\rho(y_n, f(x)) < \frac{\epsilon_o}{2} \ \forall \, n \geq n_o$ and $\forall \, x \in X$. But then this would imply that $\epsilon_o = \rho(f(x_1), f(x_2)) \leq \rho(f(x_1), y_{n_o}) + \rho(y_{n_o}, f(x_2)) < \epsilon_o$. This is a contradiction. Thus, if $x_o \in X$, then $f(x) = f(x_o) \ \forall \, x \in X$. Since $f_n \to f$, we get that $y_n \to f(x_o)$ in (Y, ρ). $\qquad \square$

Exercises

3.1.1. Let $X = (0, \infty)$ endowed with metric d' defined by: $d'(x, y) = |\frac{1}{x} - \frac{1}{y}|$ for all $x, \ y \in X$. Is (X, d') complete?

3.1.2. Show that the metric space $(\mathbb{R}, d)$ is incomplete, where $d(x, y) = |\tan^{-1} x - \tan^{-1} y|$ on $\mathbb{R}$.

3.1.3. Let $f : (X, d) \to (Y, \rho)$ be any function between two metric spaces. Then show that the metric space (X, σ), where σ is defined as

$$\sigma(x, x') = d(x, x') + \rho(f(x), f(x')) \quad \text{for} \quad x, \ x' \in X,$$

is complete if (X, d) is complete and f is continuous. (Refer to Exercise 1.3.6.)

3.1.4. Let X be a non-empty set and (Y, ρ) be a complete metric space. Then show that the metric space $B(X,Y)$ of all bounded functions from X to (Y, ρ) is complete with respect to the supremum metric.

3.1.5. Show that the space $\mathbb{R}^I$ is complete with respect to the uniform metric. (Refer to Example 1.3.9.)

3.1.6. Let $\{(X_i, d_i) : 1 \leq i \leq n\}$ be a finite collection of metric spaces. Then show that the product space $\prod_{i=1}^{n} X_i$ equipped with any of the metrics given in Example 1.3.4 is complete if and only if (X_i, d_i) is complete for all $1 \leq i \leq n$. (Keep in mind Proposition 2.6.3.) As a consequence, it can be deduced that the Euclidean space $\mathbb{R}^n$ is complete.

3.1.7. Show that the Cartesian product $\prod_{i=1}^{\infty} X_i$ is complete with respect to the metric defined in Example 1.3.10 if and only if the metric space (X_n, d_n) is complete for all $n \in \mathbb{N}$.

3.2 Cauchy Sequences *vis-à-vis* Total Boundedness

It is well known that every subsequence of a convergent sequence in a metric space (X, d) also converges to the same limit. But if a subsequence is convergent then the parent sequence need not be convergent (think of $((-1)^n)_{n \in \mathbb{N}}$). Can you think of some conditions which guarantee the convergence of the parent sequence? It is quite intuitive that if the sequential terms are arbitrarily close to each other eventually then the sequence is convergent provided it has a convergent subsequence. We are going to use this observation to prove the completeness of $(\mathbb{R}, |\cdot|)$.

Theorem 3.2.1. *The metric space $(\mathbb{R}, |\cdot|)$ is complete.*

Proof. Let (x_n) be a Cauchy sequence in $(\mathbb{R}, |\cdot|)$. First we show that (x_n) is bounded in $\mathbb{R}$. Let $\epsilon = 1$. Then there exists $n_o \in \mathbb{N}$ such that $|x_n - x_m| < 1$ for all $n, m \geq n_o$. In particular, $|x_n - x_{n_o}| < 1$ for all $n \geq n_o$, that is, $|x_n| < |x_{n_o}| + 1 \ \forall \ n \geq n_o$. Let $M = \max\{|x_1|, \ldots, |x_{n_o-1}|, |x_{n_o}| + 1\}$. Then $|x_n| \leq M$ for all $n \in \mathbb{N}$. By Bolzano–Weierstrass theorem, there exists a subsequence (x_{k_n}) of (x_n) such that $x_{k_n} \to x$ for some $x \in \mathbb{R}$. So given $\epsilon > 0$, there exists $n_\epsilon \in \mathbb{N}$ such that $|x_{k_n} - x| < \epsilon/2$ for all $n \geq n_\epsilon$. Again since (x_n) is

Cauchy in $\mathbb{R}$, there exists $n'_\epsilon \in \mathbb{N}$ such that $|x_n - x_m| < \epsilon/2$ for all $n, m \geq n'_\epsilon$. Choose $n''_\epsilon = \max\{n_\epsilon, n'_\epsilon\}$. If $n \geq n''_\epsilon$, then $k_n \geq n \geq n''_\epsilon$ and hence

$$|x_n - x| = |x_n - x_{k_n} + x_{k_n} - x|$$
$$\leq |x_n - x_{k_n}| + |x_{k_n} - x|$$
$$< \epsilon/2 + \epsilon/2 = \epsilon \ \forall \ n \geq n''_\epsilon.$$

Thus, $x_n \to x$ in $(\mathbb{R}, |\cdot|)$ and consequently, $(\mathbb{R}, |\cdot|)$ is a complete metric space. $\qquad\square$

The reader should observe carefully that in the proof of the last result, we have implicitly proved that if a Cauchy sequence (x_n) has a convergent subsequence then the sequence (x_n) itself converges to the same limit. Although it is proved in the context of real sequences but the proof can be imitated for a sequence in a general metric space as well. Further, we proved that every Cauchy sequence in $(\mathbb{R}, |\cdot|)$ is bounded. Let us prove this fact for a general metric space.

Proposition 3.2.1. *Every Cauchy sequence in a metric space (X, d) is bounded.*

Proof. Let (x_n) be a Cauchy sequence in (X, d). Then there exists $n_o \in \mathbb{N}$ such that $d(x_n, x_{n_o}) < 1 \ \forall \ n \geq n_o$. Let $M > \max\{d(x_1, x_{n_o}), d(x_2, x_{n_o}), \ldots, d(x_{n_o-1}, x_{n_o}), 1\}$. Then $(x_n) \subseteq B(x_{n_o}, M)$. $\qquad\square$

Note that if (x_n) is Cauchy, then the set $\{x_n : n \in \mathbb{N}\}$ is much stronger than being bounded. Because if (x_n) is Cauchy in (X, d), then the sequence can be contained in finitely many balls of arbitrarily small radius. Now let us generalize this notion for any arbitrary subset of a metric space.

Definition 3.2.1. Let A be a nonempty subset of a metric space (X, d). Then A is called totally bounded in (X, d) if given $\epsilon > 0$, there exists a finite subset (depending on ϵ) $\{x_1, x_2, \ldots, x_n\}$ of X such that $A \subseteq \bigcup_{i=1}^n B(x_i, \epsilon)$.

Remark. In the previous definition, the finite set $\{x_1, x_2, \ldots, x_n\}$ can be chosen from A itself (Prove it!).

The empty set $\emptyset$ is vacuously totally bounded. It is evident that a totally bounded set in a metric space is always bounded, but the converse need not be true. Think about the discrete metric. *Can you think of a bounded set in $(\mathbb{R}, |\cdot|)$ which is not totally bounded?*

If (x_n) is a Cauchy sequence in (X, d), then it is easy to see that the set $\{x_n : n \in \mathbb{N}\}$ is totally bounded. In fact, the next result talks about the relation between totally bounded sets and Cauchy sequences in a precise manner.

Theorem 3.2.2. *Let (X, d) be a metric space and A be a non-empty subset of X. Then A is totally bounded if and only if every sequence in A has a Cauchy subsequence.*

Proof. First suppose that A is totally bounded and let (x_n) be a sequence in A. Consider the set $S = \{x_n : n \in \mathbb{N}\}$. We may assume S to be infinite. Otherwise if S is finite, then (x_n) has a constant subsequence.

Now given $\epsilon = \frac{1}{2}$, there exists a finite subset F_1 of X such that $A \subseteq \bigcup_{y \in F_1} B(y, \frac{1}{2})$. Since S is infinite, there exists $y_1 \in F_1$ such that $B(y_1, \frac{1}{2})$ contains a subsequence (x_{k_n}) of (x_n) and the set $S_1 = \{x_{k_n} : n \in \mathbb{N}\}$ is infinite. Denote the subsequence (x_{k_n}) by $(x_n^{(1)})$. Now choose $\epsilon = \frac{1}{2^2}$ and repeat the previous process, that is, there exists a finite subset F_2 of X such that $A \subseteq \bigcup_{y \in F_2} B(y, \frac{1}{2^2})$. As before, there exists $y_2 \in F_2$ such that $B(y_2, \frac{1}{2^2})$ contains a subsequence of $(x_n^{(1)})$. Call this subsequence by $(x_n^{(2)})$ and note that the set $S_2 = \{x_n^{(2)} : n \in \mathbb{N}\}$ is again infinite. Proceeding in this way, we obtain a sequence of sequences, each of which is a subsequence of the preceeding one,

$$(x_n), (x_n^{(1)}), (x_n^{(2)}), \ldots, (x_n^{(k)}), \ldots$$

Note that for each $k \in \mathbb{N}$, $S_k = \{x_n^{(k)} : n \in \mathbb{N}\}$ is infinite and S_k is contained in an open ball of radius $\frac{1}{2^k}$. Now $(x_n^{(n)})$ is a subsequence of (x_n). Let $\epsilon > 0$ and choose $n_o \in \mathbb{N}$ such that $\frac{1}{2^{n_o-1}} < \epsilon$. For $n > n_o$ and $p \in \mathbb{N}$, we have

$$d(x_{n+p}^{(n+p)}, x_n^{(n)}) < \frac{1}{2^n} + \frac{1}{2^n} = \frac{1}{2^{n-1}} < \frac{1}{2^{n_o-1}} < \epsilon.$$

But this means that $(x_n^{(n)})$ is a Cauchy sequence.

Conversely, suppose that every sequence in A has a Cauchy subsequence. In order to show that A is totally bounded, choose any $\epsilon > 0$. Let $x_1 \in A$. If $A - B(x_1, \epsilon)$ is empty, then $A \subseteq B(x_1, \epsilon)$ and we are done. Otherwise, choose $x_2 \in A - B(x_1, \epsilon)$. If $A - [B(x_1, \epsilon) \cup B(x_2, \epsilon)]$ is empty, we are done. We only need to assert that this process terminates. If it does not terminate, then we would land up with a sequence (x_n) of all distinct points in A such that $d(x_i, x_j) \geq \epsilon$ for all $i \neq j$. Obviously, the sequence (x_n) cannot have a Cauchy subsequence which is contrary to our hypothesis. $\qquad\square$

Remark. In the first part of the proof, we produced a subsequence $(x_n^{(n)})$ of (x_n). Here we are employing a trick which is well-known as the diagonal method. Also satisfy yourself that in this diagonal process, you are really picking up a subsequence of (x_n) (we are moving towards right and not going back while picking up $x_n^{(n)}$).

Consequently, if A is an infinite totally bounded subset of a metric space (X, d) then there exists a Cauchy sequence of distinct points in A.

Recall that $\mathbb{R}$ is homeomorphic to the interval $(-\frac{\pi}{2}, \frac{\pi}{2})$ (Example 2.3.1). From this, it is evident that total boundedness is not preserved by homeomorphism. But interestingly, now we prove that Cauchy-continuity preserves total boundedness.

Corollary 3.2.1. *Let $f : (X, d) \to (Y, \rho)$ be a Cauchy-continuous function. If A is a totally bounded subset of (X, d), then $f(A)$ is totally bounded in (Y, ρ).*

Proof. We will use the sequential characterization of totally bounded sets to prove this. Let $(f(a_n))$ be a sequence in $f(A)$. Since A is totally bounded, (a_n) has a Cauchy subsequence say (a_{k_n}). By the Cauchy continuity of f, $(f(a_{k_n}))$ is Cauchy in (Y, ρ). Since $(f(a_n))$ was an arbitrary sequence in (Y, ρ), every sequence in $f(A)$ has a Cauchy subsequence. Thus by Theorem 3.2.2, $f(A)$ is totally bounded. $\qquad\square$

Subsequently, we can say that every Cauchy-continuous function is bounded on totally bounded subsets of the domain space. It should be noted that the converse of Corollary 3.2.1 may not hold true. For example, let $X = \{\frac{1}{n} : n \in \mathbb{N}\}$ and $\{0, 1\}$ be equipped with the metric induced by the usual distance metric. Consider the function $f : X \to \{0,1\}$ defined by: $f(x) = 0$ if n is even and $f(x) = 1$ if n is odd. Clearly, f carries totally bounded sets of X to totally bounded sets of $\{0,1\}$. In fact, every subset of $\{0,1\}$ is totally bounded. Moreover, f is continuous (as every point of X is an isolated point) but not Cauchy-continuous (take $(\frac{1}{n})$) (See Exercise 3.2.4). As a consequence of Corollary 3.2.1, it can be said that total boundedness is preserved under Cauchy equivalent metrics (Exercise 2.6.3), that is, if the metrics d and ρ on X are Cauchy equivalent then the collection of the totally bounded sets with respect to the metric d coincides with the corresponding collection in (X, ρ).

Proposition 3.2.2. *Let (X, d) be a metric space and B be a nonempty totally bounded subset of X. Then B is separable.*

Proof. Since B is totally bounded, for each $n \in \mathbb{N}$ there exists a finite subset F_n of B such that $B \subseteq \bigcup_{x \in F_n} B(x, \frac{1}{n})$. Let $F = \bigcup_{n=1}^{\infty} F_n$. Then F is countable because countable union of finite sets is countable. We claim that F is dense in B. Given $\epsilon > 0$, choose $n_o \in \mathbb{N}$ such that $\frac{1}{n_o} < \epsilon$. Let $x \in B$, then there exists $x_o \in F_{n_o}$ such that $x \in B(x_o, \frac{1}{n_o})$. Hence $x_o \in B(x, \frac{1}{n_o})$ and therefore $B(x, \epsilon) \cap F \neq \emptyset$. This implies that $x \in \overline{F}$. So $B \subseteq \overline{F}$. But F is countable. Hence B is separable. $\qquad\square$

Can you find a separable metric space which is not totally bounded? (Think of discrete metric.)

Exercises

3.2.1. Let A be a subset of a metric space (X, d). Show that if A is bounded (totally bounded) then $\overline{A}$ is also bounded (totally bounded).

3.2.2. Show that a subset of a totally bounded set in a metric space is also totally bounded.

3.2.3. Let $f : (X, d) \to (Y, \rho)$ be a function. Then prove that the following are equivalent statements:

(a) f is Cauchy-continuous.
(b) f is uniformly continuous on every totally bounded subset of X.

3.2.4. Let $f : (X, d) \to (Y, \rho)$ be a function. Then show that the following statements are equivalent:

(a) f maps totally bounded subsets of X to totally bounded subsets of Y.
(b) If (x_n) is Cauchy in X, then $(f(x_n))$ has a Cauchy subsequence in Y.

3.2.5. Show that the closed unit ball in ℓ^∞ is not totally bounded.

3.2.6. Prove that (X, d) is totally bounded if and only if $(\mathrm{CLB}(X), d_H)$ is totally bounded. (Refer to Exercise 1.5.7.)

3.3 Cantor's Intersection Theorem

Let us first explore the close connection between a complete metric space and its closed subsets. For example, consider two subsets $[0, 1]$ and $(0, 1)$ of the complete metric space $(\mathbb{R}, | \cdot |)$ with the induced metric, that is, consider the metric spaces $([0, 1], | \cdot |)$ and $((0, 1), | \cdot |)$. Then note that $([0, 1], | \cdot |)$ is complete, but $((0, 1), | \cdot |)$ is not. Why is it so? The next result answers this question.

Theorem 3.3.1. *Let (X, d) be a metric space and A be a non-empty subset of X.*

(a) *Suppose (X, d) is complete and A is closed in (X, d). Then (A, d) is a complete metric space.*
(b) *If (A, d) is a complete metric space, then A is closed in (X, d).*

Proof. (a) Let (x_n) be a Cauchy sequence in (A, d). We need to show that (x_n) converges to a point in A. Since (x_n) is also Cauchy in the complete metric space (X, d), there exists $x \in X$ such that $x_n \to x$ in (X, d). But (x_n) is a sequence from A and hence by Theorem 1.5.2(a), $x \in \overline{A}$. Since A is closed, $A = \overline{A}$ and thus $x \in A$. Therefore, the Cauchy sequence (x_n) in (A, d) converges to a point in A. Hence (A, d) is complete.

(b) Suppose (A, d) is complete. We need to show that A is closed in (X, d). Let $x \in \overline{A}$. Then by Theorem 1.5.2(a), there exists a sequence (x_n) in A such that $x_n \to x$ in (X, d). Hence (x_n) is Cauchy in (A, d). Since (A, d) is complete, there exists $y \in A$ such that $x_n \to y$ in (A, d). This means $x_n \to y$ in (X, d). Then by the uniqueness of the limit of a convergent sequence $x = y$. But $y \in A$, so $x \in A$. This implies that $\overline{A} \subseteq A$ and hence $\overline{A} = A$. So A is closed in (X, d).

$\square$

Remarks. (i) In functional analysis, two metric subspaces, c and c_0, of the metric space l^∞ play a significant role. Here, c denotes the space of all convergent sequences, while c_0 is the space of all sequences converging to zero. Clearly, $c_0 \subseteq c \subseteq l^\infty$. Moreover, these inclusions are strict. Since both these spaces are closed in the complete space l^∞ (Exercise 1.5.12), c and c_0 are also complete with respect to the supremum metric inherited from the space l^∞.

(ii) Later it will be proved that the space $C[a, b]$, equipped with the uniform metric (see Example 3.1.5), is a closed subset of $(B[a, b], D)$. Hence by Theorem 3.3.1, $(C[a, b], D)$ is complete.

Next we want to prove the famous Cantor's Intersection Theorem on a complete metric space. It is a very powerful tool which is often used to prove other important results in analysis.

Theorem 3.3.2 (Cantor's Intersection Theorem). *Let* (X, d) *be a complete metric space and let* (A_n) *be a sequence of non-empty closed subsets of* X *such that* $A_{n+1} \subseteq A_n$ $\forall$ $n \in \mathbb{N}$ *and* $\lim_{n \to \infty} d(A_n) = 0$. *Then the intersection* $\bigcap_{n=1}^{\infty} A_n$ *consists precisely of one point.*

Proof. The proof mainly consists of two parts: existence and uniqueness of the point in $\bigcap_{n=1}^{\infty} A_n$. We prove the uniqueness part first. If x, $y \in \bigcap_{n=1}^{\infty} A_n$, then x, $y \in A_n$ for all n. By definition, $d(x, y) \leq d(A_n)$ $\forall$ n. But $\lim_{n \to \infty} d(A_n) = 0$. This implies that $d(x, y) = 0$ and hence $x = y$. So the intersection $\bigcap_{n=1}^{\infty} A_n$ can have at most one element.

Now we show that $\bigcap_{n=1}^{\infty} A_n$ has a point. For each $n \in \mathbb{N}$, choose $x_n \in A_n$. Note for $m \geq n$, $A_m \subseteq A_n$ and hence $d(A_m) \leq d(A_n)$. Now given $\epsilon > 0$, there exists $n_o \in \mathbb{N}$ such that $d(A_{n_o}) < \epsilon$, because $d(A_n) \to 0$. Hence for all n, $m \geq n_o$, $d(x_n, x_m) \leq d(A_{n_o}) < \epsilon$ and so (x_n) is a Cauchy sequence in (X, d). But (X, d) is complete. Hence there exists $x \in X$ such that $x_n \to x$ in (X, d). But this means that

$(x_n)_{n \geq m} \to x$ in (X, d) for all $m \in \mathbb{N}$ and hence by Theorem 1.5.2(a), $x \in \overline{A_m}$ for all $m \in \mathbb{N}$. Since A_m is closed, this implies that $x \in A_m$ for all $m \in \mathbb{N}$. Consequently, $x \in \bigcap_{n=1}^{\infty} A_n$. $\qquad \square$

Remarks. (a) By the sequence of sets (A_n), it means that there is a function $f : \mathbb{N} \to \mathcal{P}(X)$, where $\mathcal{P}(X)$ denotes the power set of X, such that $f(n) = A_n$.

(b) In place of '$A_{n+1} \subseteq A_n$ $\forall n \in \mathbb{N}$', sometimes we call it a decreasing sequence of sets.

(c) Note that the assumption of (X, d) being a complete metric space in Cantor's Intersection Theorem cannot be dropped. For example, let $X = (0, 1)$ and consider the usual distance metric $| \cdot |$ on X. Then $(X, | \cdot |)$ is not complete. The sequence $(\frac{1}{n+1})$ does not converge in $(X, | \cdot |)$, though it is a Cauchy sequence in $(X, | \cdot |)$. Let $A_n = (0, \frac{1}{n+1}]$. Then each A_n is closed in $(X, | \cdot |)$ and $\bigcap_{n=1}^{\infty} A_n = \emptyset$.

(d) One should observe that the assumption of diameter tending to 0 is not just required for proving the uniqueness part in the theorem but also for the existence part: consider $\mathbb{N}$ with the usual distance metric and $A_n = \{k \in \mathbb{N} : k \geq n\}$. Then $d(A_n) \nrightarrow 0$ and $\bigcap_{n=1}^{\infty} A_n = \emptyset$.

The reader should observe that the converse of Cantor's Intersection Theorem is also true as proved in the next proposition.

Proposition 3.3.1. *Let* (X, d) *be a metric space such that for every sequence of non-empty closed subsets* (A_n) *in* X *with* $A_{n+1} \subseteq A_n$ $\forall n \in \mathbb{N}$ *and* $\lim_{n \to \infty} d(A_n) = 0$, *the intersection* $\bigcap_{n=1}^{\infty} A_n \neq \emptyset$. *Then the metric space* (X, d) *is complete.*

Proof. Let (x_n) be a Cauchy sequence in (X, d) and let (A_n) be a sequence of subsets of X, where $A_n = \overline{\{x_k : k \geq n\}}$ for all $n \in \mathbb{N}$. Then clearly A_n's are non-empty closed subsets of X with $A_{n+1} \subseteq A_n$ $\forall n \in \mathbb{N}$. Since (x_n) is Cauchy, $\lim_{n \to \infty} d(A_n) = 0$. Then by the given condition, let $x \in \bigcap_{n=1}^{\infty} A_n$. This implies that for every $\epsilon > 0$, $x_n \in B(x, \epsilon)$ for infinitely many n. Hence x is a cluster point of (x_n) and by Theorem 1.6.1 there exists a subsequence of (x_n) which converges to x. But since (x_n) is Cauchy, $x_n \to x$. Thus, (X, d) is complete. $\qquad \square$

Exercises

3.3.1. Show that (X, d) is complete if and only if for some fixed $r \in (0, 1)$, every sequence (x_n) in X satisfying $d(x_n, x_{n+1}) < r^n$ $\forall\, n \in \mathbb{N}$ converges in (X, d).

3.3.2 ([14]). Give an example of a non-convergent sequence (x_n) in $\mathbb{R}$ such that $|x_n - x_{n+1}|$ tends to 0 as n tends to infinity. This gives an idea of a special class of sequences which is weaker than Cauchy sequences. Let us first define them precisely: a sequence (x_n) in (X, d) is said to be *quasi-Cauchy* if given any $\epsilon > 0$, there exists n_ϵ such that $d(x_n, x_{n+1}) < \epsilon$ $\forall\, n \geq n_\epsilon$.

(a) Give an example to show that quasi-Cauchy sequences need not be bounded.
(b) Prove that a sequence (x_n) in a metric space is Cauchy if and only if every subsequence of (x_n) is quasi-Cauchy.
(c) If $f : (X, d) \to (Y, \rho)$ is uniformly continuous then show that $(f(x_n))$ is quasi-Cauchy in (Y, ρ) whenever (x_n) is quasi-Cauchy in (X, d).
(d) Let I be any interval in $\mathbb{R}$ and let (a_n) and (b_n) be sequences in I such that $|a_n - b_n| \to 0$ as $n \to \infty$. Then establish the existence of a quasi-Cauchy sequence (x_n) in I such that for every $n \in \mathbb{N}$ there exists a $k_n \in \mathbb{N}$ such that $a_n = x_{k_n}$ and $b_n = x_{k_n+1}$.
(e) Prove that a function $g : I \to \mathbb{R}$ is uniformly continuous on the interval I if and only if it preserves quasi-Cauchy sequences.

3.3.3. Use Proposition 3.3.1 to prove the completeness of $(\mathbb{R}, |\cdot|)$.

3.3.4. Prove that the space of all polynomial functions, defined on some closed and bounded interval $[a, b]$, with the uniform metric, $D(x, y) = \max_{t \in [a,b]} |x(t) - y(t)|$, is incomplete.

3.3.5. Prove the following generalization of Cantor's intersection theorem, which is well known as *Kuratowski's intersection theorem*. Let (X, d) be a metric space. Then the following are equivalent:

(1) (X, d) is complete.
(2) Whenever (A_n) is a sequence of non-empty closed subsets of X such that $A_{n+1} \subseteq A_n$ $\forall\, n \in \mathbb{N}$ and $\lim_{n \to \infty} \alpha(A_n) = 0$, then $\bigcap_{n=1}^{\infty} A_n \neq \emptyset$.

Note that for each $A \subseteq X$,

$$\alpha(A) = \inf\{\epsilon > 0 : A \text{ is contained in a finite union of balls of radius } \epsilon\}.$$

The functional α is called the *Kuratowski measure of non-compactness*. Can we further say that the intersection $\bigcap_{n=1}^{\infty} A_n$ consists precisely of one point?

3.4 Completion of a Metric Space

The very first property which we look for in any metric space is completeness. Because many good results on metric spaces only hold for complete metric spaces. Cantor's Intersection Theorem is one such result and in fact in upcoming sections also we will realize the significance of this property. So now the question is to find the 'smallest' complete metric space (Y, ρ), for an arbitrary incomplete metric space (X, d), such that $X \subseteq Y$ and $\rho|_{X \times X} = d$? Here 'smallest' means that X is somewhat dense in the complete metric space (Y, ρ). Since we are working with arbitrary metric spaces, it might not be possible to simply add the limits of the Cauchy sequences. So in a general scenario, we make use of an isometry because it preserves the metric structure.

Before we demonstrate the process of formulating such complete superspaces. We first define them formally.

Definition 3.4.1. Let (X, d) be a metric space. A completion of (X, d) is a metric space $(\widehat{X}, \widehat{d})$ with the following properties:

(a) $(\widehat{X}, \widehat{d})$ is complete.
(b) (X, d) is isometric to a dense subset of $(\widehat{X}, \widehat{d})$, that is, there exists an isometry $T : (X, d) \to (\widehat{X}, \widehat{d})$ such that $T(X)$ is dense in $(\widehat{X}, \widehat{d})$.

Remark. Since metric space properties are preserved under isometries, we can think of X and $T(X)$ to be the same. As a consequence, X itself can be treated as a dense subset of $\widehat{X}$.

Examples 3.4.1. (a) Clearly, $(\mathbb{R}, |\cdot|)$ is a completion of $(\mathbb{Q}, |\cdot|)$ and $(\mathbb{R} \setminus \mathbb{Q}, |\cdot|)$.

(b) Later, it will be proved (Weierstrass approximation theorem) that the set of all polynomials P defined on $[a, b]$ is dense in the complete metric space $(C[a, b], D)$, where D is the uniform metric. Consequently, $(C[a, b], D)$ is a completion of (P, D).

(c) Recall from Example 3.1.5 that the metric space $(C[a, b], d_p)$ is incomplete. The incompleteness of the space $(C[a, b], d_p)$ is one of the drawbacks of Riemann integration. In fact, this pushes towards a more powerful theory of integration namely 'Lebesgue integral'. One of the crucial by-product of the Lebesgue theory is L^p *space* (also called *Lebesgue space*). In fact, such spaces play a key role in measure theory and functional analysis. Interestingly, $(C[a, b], d_p)$ is dense in $L^p[a, b]$ and the space $L^p[a, b]$ is complete as well. This implies that the Lebesgue space is a completion of the space $(C[a, b], d_p)$. We are not going into the details of L^p spaces as it is out of the scope of this book.

Theorem 3.4.1. *For every metric space (X, d), there exists a completion $(\widehat{X}, \widehat{d})$.*

Proof. Recall that $B(X)$ denotes the set of all real-valued bounded functions on X. It has been shown in Example 3.1.6 that $(B(X), D)$ is a complete metric space where

$$D(f, g) = \sup_{x \in X} |f(x) - g(x)|.$$

We only need to find an isometry $T : (X, d) \to (B(X), D)$. Then the closure $\overline{T(X)}$ of $T(X)$ in $(B(X), D)$ will be the required $\widehat{X}$ (by Theorem 3.3.1) and $\widehat{d}$ will be simply D restricted to $\overline{T(X)} \times \overline{T(X)}$. Now let us start the proof.

Let $a \in X$ and keep it fixed. For each $x \in X$, define $f_x : X \to \mathbb{R}$ by $f_x(y) = d(x, y) - d(y, a) \;\forall\; y \in X$. By Proposition 1.3.1, it follows that $|f_x(y)| = |d(x, y) - d(y, a)| \leq d(x, a) \;\forall\; y \in X$. Since both x and a are fixed, $f_x \in B(X)$. Now define $T : (X, d) \to (B(X), D)$ by $T(x) = f_x$ for $x \in X$. Let us calculate $D(T(x), T(z))$ for $x,\; z \in X$.

$$|f_x(y) - f_z(y)| = |d(x, y) - d(y, a) - [d(z, y) - d(y, a)]|$$
$$= |d(x, y) - d(z, y)|$$
$$\leq d(x, z).$$

Hence $D(f_x, f_z) = \sup_{y \in X} |f_x(y) - f_z(y)| \le d(x, z)$. On the other hand,

$$|f_x(z) - f_z(z)| = |d(x, z) - d(z, z)| = d(x, z).$$

And consequently, $d(x, z) \le D(f_x, f_z)$. Hence $D(f_x, f_z) = d(x, z) \ \forall \ x, \ z \in X$. Since $(B(X), D)$ is a complete metric space, by Theorem 3.3.1 $(\overline{T(X)}, D)$ is complete. Hence by definition, $(\overline{T(X)}, D)$ is a completion of (X, d). $\qquad\square$

The proof presented here is an easy one but it is not intuitive. Let us give an outline of the original proof as well for the interested readers: we need to construct a space which contains the limits of all the non-convergent Cauchy sequences in (X, d). But there might be Cauchy sequences which converge to the same limit. We know that such sequences would be asymptotic and vice-versa (see Proposition 2.3.2 and Exercise 2.3.8). So the set of all Cauchy sequences is split into equivalence classes, where we define two Cauchy sequences (x_n) and (y_n) to be equivalent if they are asymptotic, that is,

$$(x_n) \sim (y_n) \Leftrightarrow \lim_{n \to \infty} d(x_n, y_n) = 0.$$

Verify that $\sim$ is an equivalence relation. Let us denote the set of all equivalence classes by $\widetilde{X}$. Now define a metric (check!) $\widetilde{d}$ on $\widetilde{X}$ as follows:

$$\widetilde{d}(\widetilde{x}, \widetilde{y}) = \lim_{n \to \infty} d(x_n, y_n),$$

where $\widetilde{x}, \widetilde{y} \in \widetilde{X}$ and $(x_n) \in \widetilde{x}, (y_n) \in \widetilde{y}$.

Convince yourself that the definition of $\widetilde{d}$ is well defined, that is, (i) $\lim_{n \to \infty} d(x_n, y_n)$ exists in $\mathbb{R}$ (by Proposition 2.5.1; here we are using the completeness of $(\mathbb{R}, |\cdot|)$), and (ii) the definition of $\widetilde{d}$ is independent of the choice of elements of $\widetilde{x}$ and $\widetilde{y}$. Finally, prove that the mapping $f : X \to \widetilde{X} : f(x) = \widetilde{x}$, where $\widetilde{x} \in \widetilde{X}$ such that every Cauchy sequence in $\widetilde{x}$ converges to x, is an isometry and $f(X)$ is dense in $\widetilde{X}$. This will be subsequently used to prove the completeness of $(\widetilde{X}, \widetilde{d})$.

Theorem 3.4.2. *If $(\widetilde{X}, \widetilde{d})$ and $(\widehat{X}, \widehat{d})$ are two completions of a metric space (X, d), then there exists an onto isometry $S : (\widehat{X}, \widehat{d}) \to (\widetilde{X}, \widetilde{d})$, that is, the completion of (X, d) is unique up to isometry.*

Proof. Let $T : (X, d) \to (\widehat{X}, \widehat{d})$ and $T' : (X, d) \to (\widetilde{X}, \widetilde{d})$ be isometries of X onto $T(X)$ and $T'(X)$ which are dense in $\widehat{X}$ and $\widetilde{X}$, respectively. Then $T'oT^{-1}$ is an isometry from $T(X)$ onto $T'(X)$. The claim of the theorem now follows from Theorem 2.5.2. $\square$

Remark. Due to this uniqueness (up to isometry), we call 'a' completion of a metric space to be 'the' completion.

Let us conclude this section with an interesting remark. In Example 2.6.2, it was observed that completeness is not preserved by equivalent metrics. Now the question arises: *under what conditions, does a metric space possess an equivalent metric with respect to which it is complete?* We are answering this question in the following result. For the proof, interested readers can refer to Ref. [58, Theorem 24.12].

Theorem 3.4.3. *Let* (X, d) *be a metric space. Then the following statements are equivalent:*

(a) (X, d) *is completely metrizable, that is, there exists an equivalent metric* ρ *on* X *such that* (X, ρ) *is complete.*

(b) X *can be expressed as the intersection of countably many open subsets in its completion* $(\widehat{X}, \widehat{d})$.

Consequently, every open set in $\mathbb{R}$ is completely metrizable.

Exercise

3.4.1. Prove that two metrics d and ρ on a set X are Cauchy equivalent if and only if there is a homeomorphism between their completions that fixes X. (Refer to Exercise 2.6.3.)

3.5 Extension Theorems

Many times while proving something, we need to construct a continuous function (or a function having stronger properties) on a metric space (X, d) which possesses some particular properties on a subset A of X. In such situations, there is a requirement of an extension theorem. This will help in extending a function f, which is continuous (or Cauchy continuous or uniformly continuous) on A, to a function F

which is continuous (or Cauchy continuous or uniformly continuous) on X such that $F|_A = f$ (that is, $F(a) = f(a)$ $\forall$ $a \in A$).

Example 3.5.1. Consider the function $f : (0, \infty) \to \mathbb{R}$ defined as: $f(x) = \frac{1}{x}$. Then f is continuous on $(0, \infty)$. But f cannot be extended to a continuous function on $[0, \infty)$ because $\lim_{x \to 0+} f(x)$ does not exist.

In the previous example, note that $(0, \infty)$ is not closed in $\mathbb{R}$. Now let us prove the well-known extension theorem for continuous functions which is beneficial in extending the functions continuously under certain conditions.

Theorem 3.5.1 (Tietze's extension theorem). *Let $f : A \to [0, 1]$ be a continuous function where A is a closed subset of a metric space (X, d). Then there exists a continuous function $F : X \to [0, 1]$ such that $F(a) = f(a)$ $\forall$ $a \in A$.*

Proof. Without loss of generality, we can assume that $f : A \to [1, 2]$ (otherwise we can work on $h \circ f$ where $h(x) = 1 + x$). Now define a function $F : X \to \mathbb{R}$ as follows:

$$F(x) = \left\{ \begin{array}{ll} \frac{\inf\limits_{a \in A} f(a)d(x,a)}{d(x,A)} & : x \notin A \\ f(x) & : x \in A \end{array} \right\}.$$

Since $f(A) \subseteq [1, 2]$,

$$d(x, A) \leq \inf_{a \in A} f(a)d(x, a) \leq 2d(x, A).$$

Hence $F(x) \in [1, 2]$ $\forall$ $x \in X$. Now we only need to prove that F is continuous on X which will follow from pasting lemma. Note that $X = A \cup \overline{X \setminus A}$ where both A and $\overline{X \setminus A}$ are closed sets. Only the continuity of F on $\overline{X \setminus A}$ needs to be proved.

We first prove that F is continuous on $X \setminus A$. Let $x \in X \setminus A$. Since A is closed, $X \setminus A$ is open and hence there exists $\delta > 0$ such that $B(x, \delta) \subseteq X \setminus A$. Since $x \mapsto d(x, A)$ is a continuous function (see Example 2.4.1) and $d(x, A) > 0$ $\forall$ $x \notin A$ (see Exercise 1.5.4), it is enough to prove the continuity of the function

$\widetilde{f}(u) = \inf_{a \in A} f(a) d(u, a)$ at x. So let $\epsilon > 0$ and $y \in X \setminus A$ with $d(x, y) < \epsilon/2$. If $a \in A$, then

$$d(x, a) \le d(x, y) + d(y, a) < \frac{\epsilon}{2} + d(y, a).$$

Thus, we have $f(a)d(x, a) < \epsilon + f(a)d(y, a)$ because $f(a) \le 2$. Now taking the infimum over $a \in A$, we get $\widetilde{f}(x) \le \epsilon + \widetilde{f}(y)$. Similarly, it can be proved that $\widetilde{f}(y) \le \epsilon + \widetilde{f}(x)$. Consequently, we get that whenever $d(x, y) < \min\{\delta, \frac{\epsilon}{2}\}$, $|\widetilde{f}(x) - \widetilde{f}(y)| \le \epsilon$. Therefore, $\widetilde{f}$ (and hence F) is continuous at $x \in X \setminus A$.

Now suppose $x' \in \partial A = A \cap \overline{X \setminus A}$. Let $\epsilon > 0$. Since f is continuous at x', there exists a $\delta > 0$ such that

$$|F(x') - F(a)| < \epsilon \ \forall \ a \in A \cap B(x', \delta). \ (\star)$$

Now suppose $y \in (X \setminus A) \cap B(x', \delta/4)$. We will prove that $|F(x') - F(y)| < \epsilon$. But first we claim that

$$\inf_{a \in A} f(a)d(a, y) = \inf_{a \in A \cap B(x', \delta)} f(a)d(a, y).$$

Suppose $a \in A \setminus B(x', \delta)$. Then $d(y, a) \ge d(x', a) - d(x', y) \ge \delta - \frac{\delta}{4} = \frac{3\delta}{4}$. Since $f(A) \subseteq [1, 2]$, $\inf\{f(a)d(y, a) : a \in A \setminus B(x', \delta)\} \ge \frac{3\delta}{4}$. But note that $f(x')d(x', y) \le 2d(x', y) < 2\frac{\delta}{4} < \frac{3\delta}{4}$. Therefore, $\inf_{a \in A \cap B(x', \delta)} f(a)d(a, y) < \frac{3\delta}{4}$. Now

$$\inf_{a \in A} f(a)d(a, y) = \min\left\{ \inf_{a \in A \cap B(x', \delta)} f(a)d(a, y), \inf_{a \in A \setminus B(x', \delta)} f(a)d(a, y) \right\}$$

$$= \inf_{a \in A \cap B(x', \delta)} f(a)d(a, y).$$

Similarly, it can be shown that

$$d(y, A) = \inf_{a \in A \cap B(x', \delta)} d(a, y) \ (\star\star).$$

Now

$$F(y) = \frac{\inf_{a \in A} f(a)d(y, a)}{d(y, A)} = \frac{\inf_{a \in A \cap B(x', \delta)} f(a)d(y, a)}{d(y, A)}.$$

By $(\star)$,

$$\frac{(F(x') - \epsilon) \inf\limits_{a \in A \cap B(x',\delta)} d(y,a)}{d(y,A)} < \frac{\inf\limits_{a \in A \cap B(x',\delta)} f(a)d(y,a)}{d(y,A)}$$

$$< \frac{(F(x') + \epsilon) \inf\limits_{a \in A \cap B(x',\delta)} d(y,a)}{d(y,A)}$$

Thus, by $(\star\star)$, we get

$$F(x') - \epsilon < F(y) < F(x') + \epsilon.$$

Consequently, we have $|F(x') - F(y)| < \epsilon$ for all $y \in (X \setminus A) \cap B(x', \delta/4)$. Thus, by $(\star)$, we have F to be continuous at $x' \in \partial A$. $\square$

Remark. (a) From the proof of Theorem 3.5.1, it is clear that the codomain of the given function f can be any bounded interval $[a, b]$. Subsequently, the codomain of the corresponding extension would also be the same bounded interval $[a, b]$.

(b) Suppose $f : A \to (-1, 1)$ is a continuous function where A is a closed subset of (X, d). Then by Theorem 3.5.1, there exists a continuous extension $F_1 : X \to [-1, 1]$ of f. Let $A_1 = \{x \in X : F_1(x) \in \{-1, 1\}\}$. Then A and A_1 are disjoint closed subsets of X. Hence by Urysohn's lemma (Theorem 2.2.1), there exists a continuous function $g : X \to [0, 1]$ such that $g(A_1) = 0$ and $g(A) = 1$. Now one can easily verify that the function $F : X \to (-1, 1)$ given by $F(x) = g(x).F_1(x)$ is a continuous extension of f. Note that the interval $(-1, 1)$ can be replaced by any bounded interval (a, b) throughout because the two intervals are homeomorphic.

(c) In Theorem 3.5.1, we talked about the extension of a bounded continuous function defined on a closed subset of a metric space. Interestingly, we do not require the condition of boundedness for extending the given function. Suppose $f : A \to \mathbb{R}$ is continuous where A is a closed subset of (X, d). Then $\tan^{-1} \circ f : A \to (-\frac{\pi}{2}, \frac{\pi}{2})$ is a bounded continuous function on A. Now by the previous remark, let $F : X \to (-\frac{\pi}{2}, \frac{\pi}{2})$ be the continuous extension, then $\widetilde{F} = \tan \circ F$ is the required continuous extension of f. Moreover, $\inf_{a \in A} f(a) = \inf\limits_{x \in X} \widetilde{F}(x)$ and $\sup_{a \in A} f(a) = \sup\limits_{x \in X} \widetilde{F}(x)$.

Now let us see one nice application of Tietze's extension theorem.

Corollary 3.5.1. *Let (X, d) be a metric space. Then the following statements are equivalent:*

(a) *(X, d) is complete.*
(b) *Every continuous function on (X, d) with values in an arbitrary metric space is Cauchy-continuous.*
(c) *Every real-valued continuous function on (X, d) is Cauchy-continuous.*

Proof. (a) $\Rightarrow$ (b): Let (X, d) be complete. Since continuous functions preserve convergent sequences, every continuous function on (X, d) with values in any arbitrary metric space is Cauchy-continuous.

(b) $\Rightarrow$ (c): This is immediate.

(c) $\Rightarrow$ (a): Let (x_n) be a Cauchy sequence in (X, d) which does not converge. Without loss of generality, we may assume $x_n \neq x_m$ for $n \neq m$. Since (x_n) does not converge, it has no convergent subsequence. Then by Theorem 1.6.1, (x_n) has no cluster point. Thus, $A = \{x_n : n \in \mathbb{N}\}$ is a closed subset of X by Theorem 1.4.6. Now consider the function $f : A \to \mathbb{R}$ defined as $f(x_n) = n$ for $n \in \mathbb{N}$. Then f is continuous by Theorem 2.1.1(d). Consequently, by Tietze's extension theorem there exists a real-valued continuous extension $F : (X, d) \to \mathbb{R}$ of f. This implies that $F(x_n) = f(x_n) = n$. By (c), F is Cauchy-continuous. Thus, $(F(x_n)) = (n)$ should be Cauchy in $\mathbb{R}$, which is a contradiction. $\square$

Example 3.5.1 also shows that a function f which is continuous on a dense set may not be extended continuously to the whole space. Here comes the role of uniformly continuous functions. It was seen that if f is uniformly continuous then we can surely extend the function from a dense set to the whole space provided the codomain is complete. In fact, in the next result it is proved that such a thing holds even for a weaker class as well, namely Cauchy-continuous functions. Observe that the function f in Example 3.5.1 is not Cauchy-continuous (take the sequence $(\frac{1}{n})$).

Theorem 3.5.2. *Let A be a subset of a metric space (X, d) and (Y, ρ) be a complete metric space. If $f : (A, d) \to (Y, \rho)$ is a Cauchy continuous function, then there exists a unique continuous function*

$F : \overline{A} \to Y$ such that $F(a) = f(a)$ $\forall\, a \in A$. Moreover, F is Cauchy continuous.

Proof. We need to define F on $\overline{A}$. Let $x_o \in \overline{A}$. Then by Theorem 1.5.2, there exists a sequence (a_n) in A such that $a_n \to x_o$. Consequently, the sequence $(f(a_n))$ is Cauchy in (Y, ρ) which is complete. Hence $(f(a_n))$ converges to some point in Y, let us define that point to be $F(x_o)$, that is, $F(x_o) = \lim\limits_{n \to \infty} f(a_n)$.

F is well-defined: Suppose (w_n) is another sequence in A which converges to x_o. We need to show that $\lim\limits_{n \to \infty} f(a_n) = \lim\limits_{n \to \infty} f(w_n)$. Since the sequence $\langle a_1, w_1, a_2, w_2, \ldots \rangle$ converges to x_o, its corresponding image under f is Cauchy and hence convergent in (Y, ρ). This implies that its subsequences $(f(a_n))$ and $(f(w_n))$ converge to the same limit. Hence our claim is proved.

F is Cauchy-continuous: Let (x_n) be a Cauchy sequence in $\overline{A}$. For each $n \in \mathbb{N}$, let $(a_m^n)_{m \in \mathbb{N}}$ be a sequence in A which converges to x_n. Thus, $F(x_n) = \lim\limits_{m \to \infty} f(a_m^n)$ for $n \in \mathbb{N}$. Now for each $n \in \mathbb{N}$, we can choose $m_n \in \mathbb{N}$ such that $d(a_{m_n}^n, x_n) < \frac{1}{n}$ and $\rho(f(a_{m_n}^n), F(x_n)) < \frac{1}{n}$. This implies that $(a_{m_n}^n) \asymp (x_n)$. Since (x_n) is Cauchy, by Proposition 2.3.2 the sequence $(a_{m_n}^n)$ is also Cauchy in A. Since f is Cauchy-continuous, $(f(a_{m_n}^n))$ is Cauchy. Consequently, $(F(x_n))$ is Cauchy as $(f(a_{m_n}^n)) \asymp (F(x_n))$. Therefore, F is Cauchy-continuous.

F is an extension of f: If $x_o \in A$, then we know that the constant sequence (x_o) in A converges to x_o. Thus, $F(x_o) = \lim\limits_{n \to \infty} f(x_o) = f(x_o)$. This implies that $F|_A = f$.

Uniqueness: Let $\widetilde{F} : \overline{A} \to Y$ be another continuous function such that

$$\widetilde{F}(a) = f(a) \quad \forall\, a \in A. \quad (\star)$$

Let $x \in \overline{A}$. Then there exists a sequence (a_n) in A such that $a_n \to x$. Since $\widetilde{F}$ is continuous, $\widetilde{F}(a_n) \to \widetilde{F}(x)$. By $(\star)$, $\widetilde{F}(x) = \lim\limits_{n \to \infty} \widetilde{F}(a_n) = \lim\limits_{n \to \infty} f(a_n) = F(x)$. Since x was arbitrarily chosen from $\overline{A}$, $\widetilde{F}(x) = F(x)$ $\forall\, x \in \overline{A}$. $\qquad\square$

Remark. (a) It is noteworthy that in the previous theorem if the function f is uniformly continuous then so is F (Exercise!).

(b) In Theorem 3.5.2, the condition of completeness of (Y, ρ) cannot be omitted. For example, take the identity function on $A = \{\frac{1}{n} : n \in \mathbb{N}\}$. Here $Y = A$ which is incomplete with respect to the usual distance metric and $\overline{A} = A \cup \{0\}$ in $\mathbb{R}$. Thus, the function is in fact uniformly continuous on A but it cannot be extended continuously to $\overline{A}$.

In the previous result, a Cauchy-continuous function was extended to the closure of its domain A, where $A \subseteq X$. But now the question arises, whether we can extend it to the whole of X (like we did in Tietze's extension theorem).

Theorem 3.5.3. *Let A be a subset of a metric space (X, d) and $f : (A, d) \to \mathbb{R}$ be a Cauchy-continuous function. Then there exists a real-valued Cauchy-continuous function g on (X, d) such that $g(a) = f(a) \ \forall \, a \in A$.*

Proof. Since $f : (A, d) \to \mathbb{R}$ is Cauchy-continuous and A can also be considered as a subset of the completion $(\widehat{X}, d)$ of (X, d), by Theorem 3.5.2 we can get a real-valued continuous extension F of f which is defined on the closure $cl_{\widehat{X}}(A)$ of A in $(\widehat{X}, d)$. Now by Tietze's extension theorem, we can find a continuous function $\widetilde{F} : (\widehat{X}, d) \to \mathbb{R}$ such that $\widetilde{F}(a) = F(a) \ \forall \, a \in cl_{\widehat{X}}(A)$. Note that since $(\widehat{X}, d)$ is complete, $\widetilde{F}$ is Cauchy-continuous by Corollary 3.5.1. Consequently, $\widetilde{F}|_X$ is the required function. $\qquad\square$

Remark. Note that unlike Tietze's extension theorem (for continuous functions), we don't require A to be closed in (X, d) for extending the Cauchy-continuous function f.

Since in the previous theorem A is any subset of the metric space (X, d), such an extension need not be unique. Moreover, we cannot have an analogous result for uniformly continuous functions.

Example 3.5.2. Let $A = \mathbb{N}$ and $f : A \to \mathbb{R} : f(n) = n^2 \ \forall \, n \in \mathbb{N}$. Then clearly f is uniformly continuous on A. Suppose $F : [0, \infty) \to \mathbb{R}$ is uniformly continuous such that $F|_A = f$. Then there exists a $\delta > 0$ such that $|x - y| < \delta \Rightarrow |F(x) - F(y)| < 1$. By the Archimedean property of $\mathbb{R}$, there exists $n_o \in \mathbb{N}$ such that $1/n_o < \delta$.

$$|(n_o + 1)^2 - n_o^2| = |F(n_o) - F(n_o + 1)|$$
$$\leq \left| F(n_o) - F\left(n_o + \frac{1}{n_o}\right) \right| + \left| F\left(n_o + \frac{1}{n_o}\right) \right.$$
$$\left. - F\left(n_o + \frac{2}{n_o}\right) \right| + \cdots +$$
$$+ \left| F\left(n_o + \frac{n_o - 1}{n_o}\right) - F(n_o + 1) \right|$$
$$< 1 + 1 + \cdots + 1 = n_o.$$

This implies that $n_o < -1$, which is a contradiction. Consequently, F is not uniformly continuous on $[0, \infty)$.

Observe that in Example 3.5.2, f is unbounded. Interestingly, extension result like Theorem 3.5.3 holds true for bounded uniformly continuous functions. We state the result without proof here but interested readers can follow the steps given in Exercise 3.5.5 for proving it.

Theorem 3.5.4. *Let A be a subset of a metric space (X, d) and f be a real-valued bounded uniformly continuous function on A. Then f can be extended to a bounded real-valued uniformly continuous function on (X, d). Further, if the range of f is contained in the closed interval $[a, b]$, then f has a uniformly continuous extension on X whose range is also contained in $[a, b]$.*

Historically, McShane [39] was the first to obtain explicit solution for the extension problem for uniformly continuous functions in 1934. He proved using Lipschitz extension theorem, whereas in 1990 Mandelkern [36] gave an explicit construction (the same function which we used in the proof of Tietze's extension theorem) for the problem which we outline in Exercise 3.5.5.

Towards the end of this section, let us prove McShane's extension theorem for real-valued Lipschitz functions.

Theorem 3.5.5. *Let A be a subset of a metric space (X, d) and $f : (A, d) \to \mathbb{R}$ be an M-Lipschitz function. Then there exists a real-valued M-Lipschitz function F on (X, d) such that $F(a) = f(a) \ \forall \ a \in A$.*

Proof. For each $a \in A$, let us define $f_a : (X, d) \to \mathbb{R}$ by $f_a(x) = f(a) + Md(x, a)$ for $x \in X$. Now we claim that the function F defined by

$$F(x) = \inf\{f_a(x) : a \in A\}$$

for all $x \in X$, will work for us.

F is real-valued: Suppose x is an arbitrary element of X. Let $a, a' \in A$. Then

$$f(a) - f(a') \le Md(a, a') \le Md(x, a) + Md(x, a').$$

This implies that $f(a) - Md(x, a) \le f(a') + Md(x, a') \ \forall \ a' \in A$. Thus, $f(a) - Md(x, a) \le F(x)$ $(\star)$. Therefore, $F(x) \in \mathbb{R}$ for all $x \in X$.

F is M-Lipschitz: If we prove that f_a is M-Lipschitz for all $a \in A$, then F is also M-Lipschitz by Exercise 2.4.4. So let $x, x' \in X$. Then by Proposition 1.3.1, we have

$$|f_a(x) - f_a(x')| = M|d(x, a) - d(x', a)| \le Md(x, x').$$

Thus, f_a is M-Lipschitz.

$F(a) = f(a) \ \forall \ a \in A$: Let $a_o \in A$. Then $f_{a_o}(a_o) = f(a_o)$. Thus, $F(a_o) \le f(a_o)$. Moreover, by $(\star)$ we have, $f(a_o) \le F(a_o)$. Consequently, $F(a_o) = f(a_o)$. $\qquad \square$

Remark. Similarly, one can verify that the function $G(x) = \sup_{a \in A} g_a(x)$, where $g_a(x) = f(a) - Md(x, a)$, defined on X is also an M-Lipschitz extension of f (Exercise 3.5.9).

Exercises

3.5.1. Let $f : \{a, b\} \to \{0, 1\}$ be a non-constant function where $a, b \in \mathbb{R}$. Show that there does not exist any continuous function $F : [a, b] \to \{0, 1\}$ such that $F(a) = f(a)$ and $F(b) = f(b)$. Why is this example not violating Tietze's extension theorem?

3.5.2. Let $f : A \to (a, b]$ be a continuous function where A is a closed subset of (X, d). Then prove that there exists a continuous function $F : X \to (a, b]$ such that $F(x) = f(x) \ \forall \ x \in A$.

3.5.3. Show that if the function f is uniformly continuous in Theorem 3.5.2, then the unique extension F is also uniformly continuous.

3.5.4. Give an example to show that Theorem 3.5.2 does not hold true for bounded continuous functions.

3.5.5 ([36]). If A is a subset of a metric space (X, d) and f is a real-valued bounded uniformly continuous function on A. Then prove the following statements to get a bounded real-valued uniformly continuous extension of f on (X, d):

(a) It is enough to prove the result for $f : A \to [1, 2]$.
(b) It suffices to prove for A closed.
(c) Let $F : X \to \mathbb{R}$ be defined as

$$F(x) = \left\{ \begin{array}{ll} \inf\limits_{u \in A} f(u) \frac{d(x,u)}{d(x,A)} & : x \notin A \\ f(x) & : x \in A \end{array} \right\}.$$

Then $F(x) \in [1, 2] \ \forall \ x \in X$.
(d) Let $0 < \epsilon < 1$. Choose δ such that $0 < \delta < \epsilon$ and whenever $a, \ b \in A$ with $d(a, b) < 9\delta$ then $|f(a) - f(b)| < \epsilon$.
(e) If $x \notin A$ and $y \in A$ with $d(x, y) < 3\delta$, then $|F(x) - F(y)| < 4\epsilon$.
(f) If $x, \ y \in X$ with $d(x, y) < \delta^2$, then $|F(x) - F(y)| < 9\epsilon$ (make cases: $d(x, A) < 2\delta$ and $d(x, A) \geq 2\delta$).

3.5.6. Prove that the function $f(x) = \tan x$ is not Cauchy-continuous on $\left(-\frac{\pi}{2}, \frac{\pi}{2}\right)$.

3.5.7. Show that a metric space (X, d) is complete if and only if whenever A and B are disjoint closed subsets of X, there is a Cauchy-continuous function $f : (X, d) \to \mathbb{R}$ such that $f(x) = 0 \ \forall \ x \in A$ and $f(x) = 1 \ \forall \ x \in B$. Compare this with Urysohn's Lemma.

3.5.8. Let $f : (A, d) \to I$ be a Cauchy-continuous function where A is a subset of (X, d) and I is any closed interval in $\mathbb{R}$. Prove that there exists a Cauchy-continuous function $g : (X, d) \to I$ such that $g(a) = f(a) \ \forall \ a \in A$.

3.5.9. Let A be a subset of a metric space (X, d) and $f : (A, d) \to \mathbb{R}$ be an M-Lipschitz function. Then show that the function $G(x) = \sup\limits_{a \in A} g_a(x)$, where $g_a(x) = f(a) - Md(x, a)$, defined on X is also M-Lipschitz. Moreover, $G(a) = f(a) \ \forall \ a \in A$.

3.6　Baire Category Theorem

In this section, we will prove a beautiful result, namely Baire category theorem. The significance of the Baire theorem can be seen in proving the two most powerful results in functional analysis, namely the open mapping theorem and the uniform boundedness principle. Moreover, it can also be used to prove the existence of a continuous nowhere differentiable real-valued function on $[0, 1]$. But before that let us define some required notions.

Definition 3.6.1. Let A be a subset of a metric space (X, d). If int $\overline{A} = \emptyset$, then A is called nowhere dense in X.

We would like to emphasize over here that the concepts of nowhere dense and 'everywhere' dense are not complementary. For example, the set $\mathbb{Q} \cap (0, 1)$ is neither dense in $\mathbb{R}$ nor it is nowhere dense in $\mathbb{R}$. The characterization of nowhere dense sets given in the next proposition is useful in visualizing nowhere dense sets geometrically.

Proposition 3.6.1. *Let (X, d) be a metric space and $A \subseteq X$. Then the following statements are equivalent:*

(a) *A is nowhere dense in X.*
(b) *$X - \overline{A}$ is dense in X.*
(c) *For every non-empty open set U in X, there exists a non-empty open set V in X such that $V \subseteq U$ and $A \cap V = \emptyset$.*

Proof. (a) $\Rightarrow$ (b): Let W be any non-empty open subset of X. Since int $\overline{A} = \emptyset$, W is not a subset of $\overline{A}$ and hence $W \cap (X - \overline{A}) \neq \emptyset$. Therefore, $X - \overline{A}$ is dense in X.

(b) $\Rightarrow$ (c): Let U be a non-empty open set in X. Since $X - \overline{A}$ is an open dense set in X, $U \cap (X - \overline{A})$ is a non-empty open set. Let $V = U \cap (X - \overline{A})$. Clearly, $V \subseteq U$ and $V \cap A = \emptyset$.

(c) $\Rightarrow$ (a): If possible, suppose int $\overline{A} \neq \emptyset$. Let $x \in$ int $\overline{A}$. Then there exists a nhood U of x which is contained in $\overline{A}$. By (c), there exists a non-empty open set V in X such that $V \subseteq U$ and $A \cap V = \emptyset$. Since $V \subseteq U$ and $U \subseteq \overline{A}$, $V \subseteq \overline{A}$. Moreover, since $V \neq \emptyset$, let $y \in V$. This implies that $y \in \overline{A}$. Therefore, every nhood of y intersects A. In particular, $V \cap A \neq \emptyset$. But this is a contradiction. $\qquad \square$

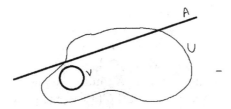

Figure 3.1 A line is nowhere dense in R^2.

From the previous proposition, it is clear that the set $\{\frac{1}{n} : n \in \mathbb{N}\}$ is nowhere dense in $(0,1)$. Also, any line is nowhere dense in $\mathbb{R}^2$ (Figure 3.1).

Definition 3.6.2. A metric space (X,d) is called a Baire space if every intersection of a non-empty countable collection of dense open subsets of X is dense in X.

Equivalently, (X,d) is a Baire space if and only if for any given countable collection $\{F_n : n \in \mathbb{N}\}$ of nowhere dense closed subsets of X, their union $\cup_{n \in \mathbb{N}} F_n$ has empty interior in X (using De Morgan's Laws and Exercise 1.4.3).

It is evident that the set of rationals $\mathbb{Q}$ is not a Baire space with respect to the usual metric: $\{\mathbb{Q} \backslash \{q\} : q \in \mathbb{Q}\}$ is a countable collection of open dense subsets in $\mathbb{Q}$ whose intersection is empty. Next is the main result of this section which gives a sufficient condition for a metric space to be a Baire space.

Theorem 3.6.1 (Baire category theorem). *Every complete metric space (X,d) is a Baire space.*

Proof. Let (A_n) be a sequence of open and dense subsets of X. We need to show that $\bigcap_{n=1}^{\infty} A_n$ is dense in (X,d). Let $x \in X$ and $r > 0$. Since A_1 is dense in (X,d), $B(x,r) \cap A_1 \neq \emptyset$. Now A_1 is open, hence $B(x,r) \cap A_1$ is open in (X,d). Let $x_1 \in B(x,r) \cap A_1$. Then there exists $r_1 > 0$ such that $r_1 \leq 1$ and $C(x_1, r_1) \subseteq B(x,r) \cap A_1$. Again since A_2 is open and dense in (X,d), $B(x_1, r_1) \cap A_2$ is open and non-empty. So as before, we may choose $x_2 \in X$ such that $x_2 \in C(x_2, r_2) \subseteq B(x_1, r_1) \cap A_2$ and $0 < r_2 \leq \frac{1}{2}$. We continue this process inductively. Suppose $x_1, \ldots, x_n$ in X and $r_1, \ldots, r_n$ have been chosen such that $x_{i+1} \in C(x_{i+1}, r_{i+1}) \subseteq B(x_i, r_i) \cap A_{i+1}$ for $1 \leq i \leq n-1$ and $0 < r_i \leq \frac{1}{i}$ for $1 \leq i \leq n$. Now consider $B(x_n, r_n) \cap A_{n+1}$ which is open and non-empty. So we may choose x_{n+1} in X such that $x_{n+1} \in C(x_{n+1}, r_{n+1}) \subseteq B(x_n, r_n) \cap A_{n+1}$ and $0 < r_{n+1} \leq \frac{1}{n+1}$. Thus, there exists a sequence (x_n) in X and a sequence (r_n) of positive numbers such that $0 < r_n \leq \frac{1}{n}$ and $C(x_{n+1}, r_{n+1}) \subseteq B(x_n, r_n) \cap A_{n+1}$ for each n. Let $C_n = C(x_n, r_n)$ for $n \in \mathbb{N}$. Then each C_n is non-empty, closed and $C_{n+1} = C(x_{n+1}, r_{n+1}) \subseteq B(x_n, r_n) \cap A_{n+1} \subseteq C(x_n, r_n) = C_n$ for each n. Moreover, $d(C_n) \leq 2r_n \leq \frac{2}{n} \Rightarrow d(C_n) \to 0$ as $n \to \infty$. But (X, d) is a complete metric space. Hence by Cantor's Intersection Theorem, there exists $y \in \bigcap_{n=1}^{\infty} C_n$. Now $y \in C_1 = C(x_1, r_1) \Rightarrow y \in B(x, r)$ and $y \in A_1$. Also $y \in C_{n+1}$ for each n which implies $y \in B(x_n, r_n) \cap A_{n+1}$ for each n. Hence $y \in \bigcap_{n=1}^{\infty} A_n$, that is, $y \in B(x, r) \cap (\bigcap_{n=1}^{\infty} A_n)$. But x and r were arbitrarily chosen. Hence $\bigcap_{n=1}^{\infty} A_n$ is dense in (X, d) and consequently (X, d) is a Baire space. $\square$

Now the curious readers might be thinking: why is the word 'category' used in the name of the theorem? So here we would like to mention that many authors first define the sets of first category and that of second category, and then they state the Baire category theorem in terms of second category spaces. Since this word 'category' has no particular significance, we refrain ourselves from defining unnecessarily extra terminologies in this book. It should be noted that often we call the previous theorem as Baire category theorem. But in practice, we mainly use the following corollary. Subsequently, this corollary is also known as Baire category theorem and it easily follows from the discussion after the definition of a Baire space.

Corollary 3.6.1. *If (X, d) is a complete metric space and $X = \bigcup_{n=1}^{\infty} A_n$, then int $\overline{A_{n_o}} \neq \emptyset$ for some $n_o \in \mathbb{N}$.*

Consequently, the set of irrationals cannot be expressed as a countable union of nowhere dense sets in $\mathbb{R}$ as $\mathbb{R}$ is complete and $\mathbb{Q} = \bigcup_{q \in \mathbb{Q}} \{q\}$. Let us now prove a nice application of Baire category theorem which is an analogue of the *uniform boundedness principle* in functional analysis.

Theorem 3.6.2. *Let F be a family of real-valued continuous functions defined on a complete metric space (X, d). Suppose that for* *every $x \in X$, there exists $M_x > 0$ such that $|f(x)| \leq M_x$ for all $f \in F$. Then there exist a non-empty open set U in X and $M > 0$ such that $|f(x)| \leq M$ for all $f \in F$ and $x \in U$.*

Proof. For $n \in \mathbb{N}$ and $f \in F$, let $A_{n,f} = \{x \in X : |f(x)| \leq n\}$. Since every $f \in F$ is continuous, $A_{n,f}$ is closed in X. Consequently, the set $A_n = \bigcap_{f \in F} A_{n,f}$ is also closed in X. Since for every $x \in X$, there exists $M_x > 0$ such that $|f(x)| \leq M_x$ for all $f \in F$, we have $X = \bigcup_{n \in \mathbb{N}} A_n$. By Corollary 3.6.1, $\text{int} A_{n_o} \neq \emptyset$ for some $n_o \in \mathbb{N}$. Thus, there exists an open set U which is contained in A_{n_o}. Hence, $|f(x)| \leq n_o$ for every $x \in U$ and $f \in F$. $\square$

Historical Perspective

In 1897, the American mathematician Osgood showed that any countable intersection of dense open sets in $\mathbb{R}$ is again dense in $\mathbb{R}$. In 1899, Baire proved this result for the Euclidean space $\mathbb{R}^n$. In 1914, Hausdorff went one step further and proved that every completely metrizable space (X, τ) has the property that any countable intersection of dense open sets in the topological space (X, τ) is again dense in (X, τ) — its metric version is proved in Theorem 3.6.1. This result eventually came to be known as the Baire category theorem. The term 'Baire space' was introduced by Bourbaki. Strictly speaking, Theorem 3.6.1 should be called Osgood–Baire–Hausdorff Theorem.

We would like to end this section with the historical perspective behind the following famous result of Stefan Banach. We do not give its proof, but would like to emphasize that Baire category theorem is needed to prove this result. Note that $(C[0,1], D)$ is a complete metric space and hence a Baire space. Here D denotes the supremum metric on $C[0,1]$ defined by $D(f,g) = \sup_{x \in [0,1]} |f(x) - g(x)|$. Of course, we also require the fact that every real-valued continuous function on $[0,1]$ is bounded (it follows from the compactness of $([0,1], |\cdot|)$ which will be discussed in the next chapter).

Theorem 3.6.3 (S. Banach, 1931). *Let A be the family of all members f of $C[0,1]$ having a derivative at some point of $[0,1]$, that is, $A = \{f \in C[0,1] : f \text{ has a derivative at some point of } [0,1]\}$. Then A can be written as a countable union of nowhere dense* *sets in $(C[0,1], D)$ and consequently, $C[0,1] \setminus A = \{f \in C[0,1] :$ f has derivative at no point of $[0,1]\}$ cannot be expressed as a union of countable collection of nowhere dense sets in $C[0,1]$ (Intuitively, $C[0,1] \setminus A$ is a large set).*

In 1834, Bernhard Bolzano gave an example of a real-valued continuous function on an interval which is nowhere differentiable in that interval. Weierstrass, though usually given credit for what Bolzano found, presented such an example of a function in lectures in 1861 and in a paper to the Berlin Academy in 1872. But for almost a century after 1834, mathematicians treated such functions as pathological. As stated before, in 1931 Banach showed that, in some sense, the vast majority of real-valued continuous functions defined on a closed and bounded interval of $\mathbb{R}$ are not differentiable anywhere.

Exercises

3.6.1. Show that the set of irrational numbers, $\mathbb{P}$, is not a countable union of closed sets in $(\mathbb{R}, |\cdot|)$. And hence $\mathbb{Q}$, the set of rationals, cannot be written as a countable intersection of open sets in $(\mathbb{R}, |\cdot|)$.

3.6.2. Show that the space of irrationals, $\mathbb{P}$, is a Baire space. This shows that the converse of the Baire category theorem is not true in general.

3.6.3. Does there exist a metric d on $\mathbb{Q}$ which is equivalent to the usual metric and $(\mathbb{Q}, d)$ is complete?

3.6.4. Show that $\mathbb{R}$ is not countable. Further, prove that if a complete metric space is countable then it must have an isolated point.

3.6.5. Let (X, d) be a metric space, Y be a non-empty subset of X and $\emptyset \neq N \subseteq Y$.

(i) If N is nowhere dense in Y, then show that N is also nowhere dense in X. Give an example to show that the converse need not hold true, that is, if N is nowhere dense in X, then it need not be nowhere dense in Y.

(ii) If Y is dense in X and N is nowhere dense in X, then prove that N is nowhere dense in Y.

3.6.6. Show that a finite union of nowhere dense sets is again a nowhere dense set. Is this statement true for a countable union of nowhere dense sets?

3.6.7. Show that a metric space (X, d) is a Baire space if and only if every non-empty open subset of X cannot be written as a countable union of nowhere dense sets in X.

3.6.8. Show that every open subset of a Baire space is itself a Baire space. Hence deduce that the unit interval $(0, 1)$ is a Baire space.

3.7 Oscillation of a Function

Recall that a finite union of closed sets is closed, while a finite intersection of open sets is open. But the same is not true in case we take countably infinite unions and countably infinite intersections respectively. Since these countable versions play a crucial role in the study, let us give some name to these special sets.

Definitions 3.7.1. Let (X, d) be a metric space.

(a) A subset of X is called a G_δ-set if it is the intersection of countably many open sets.
(b) A subset of X is called a F_σ-set if it is the union of countably many closed sets.

It is evident that G_δ-sets and F_σ-sets are complements of each other. In other words, the complement of an F_σ-set is a G_δ-set and vice versa. If we take the set of all rationals, then it is clearly an F_σ-set in $\mathbb{R}$ being a countable union of singletons which are closed in $\mathbb{R}$. And hence the set of irrationals is a G_δ-set. Moreover, the closed interval $[a, b]$ is a G_δ-set because $[a, b] = \bigcap_{n \in \mathbb{N}} (a - \frac{1}{n}, b + \frac{1}{n})$. (See Exercise 3.7.3.)

Now let us define a notion which is quite helpful in examining the set of points of continuity of a function. Let f be a function from (X, d) to (Y, ρ) and $x \in X$. For $n \in \mathbb{N}$, let

$$\omega_n(f, x) = \text{diam}(f(B(x, 1/n)))$$
$$= \sup\{\rho(f(x'), f(x'')) : x', x'' \in B(x, 1/n)\}.$$

Note that $\omega_n(f, x) \geq \omega_{n+1}(f, x)$ for each $n \in \mathbb{N}$. Thus, it is a decreasing sequence of extended real numbers which is bounded below by 0. As a consequence, $\lim\limits_{n \to \infty} \omega_n(f, x) = \inf_{n \in \mathbb{N}} \omega_n(f, x)$ is also a non-negative extended real number. This limit plays a significant role in the study of continuous functions and hence given a special name.

Definition 3.7.2. Let $f : (X, d) \to (Y, \rho)$ be a function and $x \in X$. Then the oscillation of f at x, denoted by $\omega(f, x)$, is defined as

$$\omega(f, x) = \lim_{n \to \infty} \omega_n(f, x) = \inf_{n \in \mathbb{N}} \sup_{x', x'' \in B(x, 1/n)} \rho(f(x'), f(x'')).$$

Roughly speaking, the oscillation of a function quantifies the variation in the function's extreme values in the neighbourhoods of the point x when we approach towards x.

Remarks. (a) $\omega(f, x)$ is a non-negative extended real number.
(b) If x is an isolated point of X, then $\omega(f, x) = 0$.

Let us look at an example. Consider the function $f : [-\frac{\pi}{2}, \frac{\pi}{2}] \to \mathbb{R}$ defined as $f(x) = \tan x$ if $x \in (-\frac{\pi}{2}, \frac{\pi}{2})$ and $f(-\frac{\pi}{2}) = 0 = f(\frac{\pi}{2})$. Then $\omega(f, x) = \infty$ if $x = -\frac{\pi}{2}$ or $\frac{\pi}{2}$. But $\omega(f, x) = 0$ for $x \in (-\frac{\pi}{2}, \frac{\pi}{2})$. Here observe that the function f is continuous on $(-\frac{\pi}{2}, \frac{\pi}{2})$. In fact, this can be generalized as follows.

Theorem 3.7.1. *Let $f : (X, d) \to (Y, \rho)$ be a function. Then f is continuous at a point $x \in X$ if and only if $\omega(f, x) = 0$.*

Proof. Let f be continuous at $x \in X$ and $\epsilon > 0$. Then there exists a $\delta > 0$ such that whenever $x' \in B(x, \delta)$, $\rho(f(x), f(x')) < \frac{\epsilon}{2}$. Now by Archimedean property of $\mathbb{R}$, $\frac{1}{n_o} < \delta$ for some $n_o \in \mathbb{N}$. Thus, $\omega_n(f, x) \leq \epsilon$ for all $n \geq n_o$ which implies that $\omega(f, x) = \inf_{n \in \mathbb{N}} \omega_n(f, x) \leq \epsilon$. Since $\epsilon > 0$ was arbitrarily chosen and $\omega(f, x) \geq 0$, we get $\omega(f, x) = 0$.

For the converse, let $\omega(f, x) = 0$ and $\epsilon > 0$. Then $\inf_{n \in \mathbb{N}} \omega_n(f, x) = 0 < \epsilon$. Now by the definition of infimum, there exists $n_o \in \mathbb{N}$ such that $\omega_{n_o}(f, x) < \epsilon$. Consequently, $\rho(f(x'), f(x)) < \epsilon$ for all $x' \in B(x, 1/n_o)$. Hence f is continuous at x. $\qquad\square$

Corollary 3.7.1. *Let (X, d) and (Y, ρ) be two metric spaces. Suppose A denotes the set of points of discontinuity of a function $f : (X, d) \to (Y, \rho)$. Then*

(a) *A is an F_σ-set.*
(b) *either A contains a non-empty open set or it can be expressed as a countable union of nowhere dense sets.*

Proof. (a) By Theorem 3.7.1,

$$A = \{x \in X : \omega(f, x) > 0\} = \bigcup_{n \in \mathbb{N}} \left\{x \in X : \omega(f, x) \geq \frac{1}{n}\right\}.$$

Now we just need to prove that the sets $A_n = \left\{x \in X : \omega(f, x) \geq \frac{1}{n}\right\}$ are closed in (X, d). Let $x \notin A_n$. This implies that $\omega(f, x) < \frac{1}{n}$. By the definition of infimum, there exists $n_o \in \mathbb{N}$ such that $\omega_{n_o}(f, x) < \frac{1}{n}$. Thus $\rho(f(x'), f(x'')) < \frac{1}{n}$ for all x', $x'' \in B(x, 1/n_o)$. Now if $x' \in B(x, 1/n_o)$, then we know that there exists $n_1 \in \mathbb{N}$ such that $B(x', 1/n_1) \subseteq B(x, 1/n_o)$. Consequently, $\omega_{n_1}(f, x') < \frac{1}{n}$. Hence $\omega(f, x') < \frac{1}{n}$. Thus, $x' \notin A_n$. Since x' was arbitrary element of $B(x, 1/n_o)$, we get $B(x, 1/n_o) \subseteq A_n^c$. Thus A_n^c is open and hence A_n is closed in (X, d).

(b) By (a), $A = \bigcup_{n \in \mathbb{N}} A_n$, where A_n's are closed in X. Suppose A does not contain any non-empty open set, then int $A_n = \emptyset \ \forall \ n \in \mathbb{N}$. Hence A_n is nowhere dense for all $n \in \mathbb{N}$. $\square$

Remark. The set of points of continuity of f is a G_δ-set.

Exercises

3.7.1. Compute the oscillation function for the following functions on $\mathbb{R}$:

(a) $f(x) = 1$ if x is rational and $f(x) = 0$ otherwise (*Dirichlet function*).
(b) $g(x) = 1/x$ for $x \neq 0$ and $g(0) = 0$.
(c) $h(x) = x$ if x is rational and $h(x) = 0$ otherwise.

3.7.2. (a) Show that there does not exist any function $f : [0, 1] \to \mathbb{R}$ which is continuous exactly on the rationals.
(b) Give an example of a real-valued function which is discontinuous exactly on the rationals.

3.7.3. Prove that every open set in a metric space is an F_σ-set, while every closed set is a G_δ-set.

3.7.4. Let $f : (X, d) \to \mathbb{R}$ be a bounded function. Then show that the real-valued oscillation function $\omega(f, \cdot)$ is upper semi-continuous at every point of X. Can you think of an example of a function f such that $\omega(f, \cdot)$ is not lower semi-continuous? (Refer to Exercise 2.1.9 for definitions.)

Chapter 4

Compactness

The notion of compactness is an extremely useful concept both in analysis and topology. In fact, this notion can be thought of as a generalization of finiteness in some sense. In real analysis, many of the important results implicitly use the compactness of closed and bounded intervals in $\mathbb{R}$. We are going to explore this feature in detail in the present chapter. We advice the readers to look at the paper [27] by Hewitt who was one of the great mathematicians of 20th century. He was a topologist as well as an analyst.

4.1 Basic Properties

In order to define compactness, we need to start with covers.

Definitions 4.1.1. A family $\mathcal{A} = \{A_i : i \in I\}$ of subsets of a non-empty set X is said to **cover** a subset G of X if $G \subseteq \bigcup_{i \in I} A_i$. If a subfamily of $\mathcal{A}$ covers G, then it is called a **subcover**. If (X, d) is a metric space and each A_i is open in (X, d), then $\mathcal{A}$ is called an **open cover** for G in (X, d).

Example 4.1.1. Let $X = \mathbb{R}$ and $A_n = (-n, n)$ for $n \in \mathbb{N}$. Then $\{A_n : n \in \mathbb{N}\}$ is an open cover for $\mathbb{R}$. But if we take $B_n = [-n, n)$ for each $n \in \mathbb{N}$, then $\{B_n : n \in \mathbb{N}\}$ is not an open cover for $\mathbb{R}$.

Definition 4.1.2. Let (X, d) be a metric space and $K \subseteq X$. Then K is called **compact** in (X, d) if every open cover of K has a finite subcover.

It is evident from the definition of open sets that every open set in a metric space can be written as a union of open balls. Now this observation can be applied to verify that K is compact in (X, d) if and only if every cover of K by open balls in (X, d) has a finite subcover.

Example 4.1.2. If (X, d) is a metric space, then the empty subset and any finite subset of X are (trivially) compact in (X, d). But a compact subset of X need not be finite.

Example 4.1.3. Let (X, d) be a metric space and (x_n) be a sequence in X converging to a point x in X. Then the set $A = \{x_n : n \in \mathbb{N}\} \cup \{x\}$ is compact in (X, d): let $\{U_i : i \in I\}$ be an open cover of A in (X, d). Since $A \subseteq \bigcup_{i \in I} U_i$, there exists $i_o \in I$ such that $x \in U_{i_o}$. Now $x_n \to x$ implies that there exists $n_o \in \mathbb{N}$ such that $x_n \in U_{i_o} \ \forall \, n \geq n_o$. Since $A \subseteq \bigcup_{i \in I} U_i$, for each x_j $(1 \leq j \leq n_o - 1)$ choose U_{k_j} $(k_j \in I)$ such that $x_j \in U_{k_j}$. Then $\{U_{i_o}, U_{k_1}, U_{k_2}, \ldots, U_{k_{n_o-1}}\}$ is a finite subcover for A. Hence A is compact in (X, d). In particular, consider the space $(\mathbb{R}, |\cdot|)$ and $A = \{\frac{1}{n} : n \in \mathbb{N}\} \cup \{0\}$. Then A is compact in $(\mathbb{R}, |\cdot|)$, but A is not finite.

Example 4.1.4. Let (X, d) be a metric space in which every subset of X is open. Now let A be a non-empty compact set in (X, d). Since $\{x\}$ is open in (X, d) for all $x \in X$ and $A = \bigcup_{x \in A} \{x\}$, the collection $\{\{x\} : x \in A\}$ is an open cover of A in (X, d). Now given that A is compact in (X, d), there exist $x_1, \ldots, x_n$ in A such that $A \subseteq \bigcup_{i=1}^{n} \{x_i\}$. Hence $A = \{x_1, \ldots, x_n\}$, that is, A is finite. Thus, in a metric space (X, d) in which every point is isolated, every compact subset of X must be finite. *What about the converse?* (See Exercise 4.1.1.)

Example 4.1.5. The open interval $(0, 1)$ is not compact in $(\mathbb{R}, |\cdot|)$. Consider the open cover $\mathcal{A} = \{(\frac{1}{n}, 1) : n \in \mathbb{N}\}$ for $(0, 1)$. This cover does not have any finite subcover for $(0, 1)$. Because any finite subcover of $\mathcal{A}$ will look like $\{(\frac{1}{n_k}, 1) : 1 \leq k \leq m\}$ for some $m \in \mathbb{N}$. But this will only cover the interval $(\frac{1}{n_o}, 1)$ where $n_o = \max\{n_k : 1 \leq k \leq m\}$.

Similarly, $\mathbb{R}$ itself is not compact in $(\mathbb{R}, |\cdot|)$, since $\{(-n, n) : n \in \mathbb{N}\}$ is an open cover of $\mathbb{R}$ without having any finite subcover for $\mathbb{R}$.

Interestingly, in Section 4.4 it will be proved that every closed and bounded subset of $\mathbb{R}$ is compact. In particular, any interval of the type $[a, b]$ is also compact in $\mathbb{R}$. But before that, let us discuss some elementary results on compactness.

Proposition 4.1.1. *If (X, d) is a metric space and $\emptyset \neq A \subseteq Y \subseteq X$. Then A is compact in (Y, d) if and only if A is compact in (X, d).*

Proof. Use Proposition 1.4.8. □

As a consequence of the previous proposition, it can be said that a subset A of (X, d) is compact if and only if the metric space (A, d) is compact. Thus, while talking about the compactness of a set, we do not require to mention the ambient space. Recall that in Example 4.1.5, it was seen that the open interval $(0, 1)$ is not compact in $\mathbb{R}$. The next result provides the justification for a general case.

Proposition 4.1.2. *Let (X, d) be a metric space and K be a non-empty compact subset of X. Then K is closed and bounded in (X, d).*

Proof. We first show that any point lying outside K is not a closure point of K. Pick $a \in X - K$. Now choose any $x \in K$. Since $a \neq x$, there exist open balls U_x and V_x in (X, d) such that $a \in U_x$, $x \in V_x$ and $U_x \cap V_x = \emptyset$. Now vary x over K. Then $\{V_x : x \in K\}$ is an open cover of K in (X, d). Since K is compact in (X, d), there exist $x_1, x_2, \ldots, x_n$ in K such that $K \subseteq \bigcup_{i=1}^{n} V_{x_i}$. Now consider the set $U = \bigcap_{i=1}^{n} U_{x_i}$. By construction, $a \in U_x \ \forall \ x \in K$. Hence $a \in U$. Also U, being a finite intersection of open sets in (X, d), is itself open in (X, d). So U is a nhood of a in (X, d). Now

$$U \cap \left(\bigcup_{i=1}^{n} V_{x_i} \right) = \bigcup_{i=1}^{n} (U \cap V_{x_i}) \subseteq \bigcup_{i=1}^{n} (U_{x_i} \cap V_{x_i}) = \emptyset.$$

But $K \subseteq \bigcup_{i=1}^{n} V_{x_i}$, hence $U \cap K = \emptyset$. Since U is a nhood of a in (X, d), a cannot be a closure point of K. So $\overline{K} \subseteq K$. But K is always contained in $\overline{K}$. Hence $K = \overline{K}$ and consequently K is closed in (X, d).

For proving boundedness, choose a point $x_o \in X$. Now for any $x \in K$, there exists an $n_x \in \mathbb{N}$ (depending on x) such that $d(x, x_o) < n_x$, that is, $x \in B(x_o, n_x)$. This means that $\{B(x_o, n) : n \in \mathbb{N}\}$ is an open cover of K. Since K is compact in (X, d), there exist $n_1, \ldots, n_t$ in $\mathbb{N}$ such that $K \subseteq \bigcup_{i=1}^{t} B(x_o, n_i) = B(x_o, m)$ where $m = \max_{1 \leq i \leq t} n_i$. Hence K is bounded in (X, d). $\qquad\square$

Observe carefully the need of compactness of K in the proof of Proposition 4.1.2. We could take U to be $\bigcap_{x \in K} U_x$, but then U may not be open. Here note that compactness is turned into finiteness. This proof validates the observation in the beginning of the chapter: Compactness, in some sense, is a topological generalization of finiteness.

Corollary 4.1.1. *Let (X, d) be a metric space and $x_o \in X$. If K is a compact subset of X and each point of $K \setminus \{x_o\}$ is an isolated point of K, then K is at most countable.*

Proof. If K is finite then we are done. Suppose K is infinite. Since every point of $K \setminus \{x_o\}$ is isolated, for every $x \in K \setminus \{x_o\}$, there exists $\epsilon_x > 0$ such that $B(x, \epsilon_x) \cap K = \{x\}$. Thus $x_o \in K$ otherwise $\{B(x, \epsilon_x) : x \in K \setminus \{x_o\}\}$ is an open cover of K with no finite subcover.

Since K is compact, it is bounded. Let $K \subseteq B(x_o, n_o)$. Now $U_1 = B(x_o, 1) \cup (\bigcup \{B(x, \epsilon_x) : x \in K \setminus \{x_o\}\})$ is an open cover of K and hence it has a finite subcover. This implies that there exists at most finitely many points of K, say $x_1, \ldots, x_n$, outside $B(x_o, 1)$. Similarly, we can see that $K \cap [B(x_o, \frac{1}{n}) \setminus B(x_o, \frac{1}{n+1})]$ is finite. Now $K = \bigcup_{n \in \mathbb{N}} \{K \cap [B(x_o, \frac{1}{n}) \setminus B(x_o, \frac{1}{n+1})]\} \cup \{x_1, \ldots, x_n\}$. Hence K is countably infinite. $\qquad\square$

Note that the proof of Proposition 4.1.2 actually gives the following result (also see Corollary 2.2.2).

Proposition 4.1.3. *Suppose K is compact in a metric space (X, d) and $x \in X - K$. Then there exist open sets V and W in (X, d) such that $x \in V$, $K \subseteq W$ and $V \cap W = \emptyset$ (and hence $x \notin \overline{W}$).*

Proposition 4.1.4. *Let (X, d) be a metric space. Suppose K is compact in (X, d) and F is closed in (X, d). Then F is compact if $F \subseteq K$.*

Proof. Let $\{U_i : i \in I\}$ be an open cover of F in (X, d). Since F is closed in (X, d), $X - F$ is open in (X, d). Now

$$F \subseteq \bigcup_{i \in I} U_i \Rightarrow X = (X - F) \cup F \subseteq (X - F) \cup \left(\bigcup_{i \in I} U_i \right).$$

In particular, $(X - F) \cup \{U_i : i \in I\}$ is an open cover of the compact set K. Hence, there exist $i_1, \ldots, i_n$ in I such that $K \subseteq (X - F) \cup (\bigcup_{k=1}^{n} U_{i_k})$. Note that $F \subseteq K$ and the set $X - F$ does not cover any part of F. Hence $F \subseteq \bigcup_{k=1}^{n} U_{i_k}$. This shows that $\{U_{i_k} : 1 \le k \le n\}$ is a finite subcover of the open cover $\{U_i : i \in I\}$ for F. Hence F is compact. $\qquad\square$

Do you think that the converse of Proposition 4.1.4 holds true? (See Exercise 4.1.3.)

Exercises

4.1.1. Prove that the following statements are equivalent for a metric space (X, d):

(a) Every point of (X, d) is an isolated point.
(b) Every compact subset of X is finite.

4.1.2. Give an example to show that the converse of Proposition 4.1.2 need not hold true in general.

4.1.3. If every proper closed subset of a metric space (X, d) is compact, does that imply (X, d) is compact?

4.1.4. Let $F = \{(x_n) \in \ell^\infty(\mathbb{R}) : \sup_{n \in \mathbb{N}} |x_n| \le 1\}$. Being a closed unit ball, F is closed and bounded in $(\ell^\infty(\mathbb{R}), D)$, but show that F is not compact in $(\ell^\infty(\mathbb{R}), D)$. Here D is the supremum metric on $\ell^\infty(\mathbb{R})$. (Refer to Example 1.3.8.)

4.1.5. A family of sets $\mathcal{A}$ is said to have the *finite intersection property* if whenever $A_1, A_2, \ldots, A_n \in \mathcal{A}$, $\bigcap_{i=1}^{n} A_i \ne \emptyset$.
    Prove that a metric space (X, d) is compact if and only if every family of closed sets with the finite intersection property has a non-empty intersection.

4.1.6. Show that intersection (union) of an arbitrary (finite) collection of non-empty compact subsets of a metric space (X, d) is also compact. What about countable union of compact sets?

4.1.7. (1) Let A be a compact subset of the Euclidean space $\mathbb{R}^n$ $(n > 1)$ with countable boundary ∂A. Then prove that $A = \partial A$.
(2) Give an example to show that (i) need not hold true for $n = 1$.
(3) If A is a compact subset of $\mathbb{C}$ with countable boundary, then prove that $A = \partial A$.

4.2   Equivalent Characterizations of Compactness

In this section, we essentially look at two nice characterizations of compact subsets of a metric space: one in terms of sequences and the other one in terms of accumulation points. The sequential characterization of compact subsets of a metric space is a very powerful and useful tool. In order to present these characterizations in a simpler way, we need to discuss some preliminary results.

Proposition 4.2.1. *Let (X, d) be a metric space and A be a non-empty compact subset of (X, d). Then A is totally bounded in (X, d).*

Proof. Let $r > 0$. Then $\{B(x, r) : x \in A\}$ is an open cover of A in (X, d). Since A is compact in (X, d), there exist $x_1, \ldots, x_n$ in A such that $A \subseteq \bigcup_{i=1}^{n} B(x_i, r)$. Thus, A is totally bounded in (X, d).   □

As a consequence of the previous proposition, we can see that the closed unit ball $B[0, 1]$ in l^2-space is not compact: note that $d(e_n, e_m) = \sqrt{2} \ \forall \ m \neq n$, where $e_n \in l^2$ is the sequence whose nth term is 1 and rest all are 0. Thus, the sequence (e_n) in $B[0, 1]$ has no Cauchy subsequence. Now see Theorem 3.2.3.

The next corollary immediately follows from Proposition 3.2.5.

Corollary 4.2.1. *Every compact metric space (X, d) is separable.*

Definition 4.2.1. Let A be a subset of a metric space (X, d) and $\mathcal{U} = \{U_i : i \in I\}$ be an open cover of A in (X, d). Suppose that there exists some $\delta > 0$ such that for each $x \in A$, we have $B(x, \delta) \subseteq U_i$ for some $i \in I$. Then δ is called a **Lebesgue number** of the cover $\mathcal{U}$.

Proposition 4.2.2. *Let (X, d) be a metric space and A be a non-empty subset of X. Suppose that every sequence (x_n) in A has a subsequence (x_{k_n}) converging to a point in A. Then every open cover of A in (X, d) has a Lebesgue number.*

Proof. Let $\mathcal{U} = \{U_i : i \in I\}$ be an open cover of A in (X, d). If possible, suppose that for this open cover $\mathcal{U}$, there does not exist any Lebesgue number. So for each $n \in \mathbb{N}$, there exists $x_n \in A$ such that $B(x_n, \frac{1}{n}) \not\subseteq U_i \ \forall \ i \in I$, that is, $B(x_n, \frac{1}{n}) \cap (X - U_i) \neq \emptyset$ for each $i \in I$. $(\star)$

Now by the given condition, there exists a subsequence (x_{k_n}) of (x_n) such that $x_{k_n} \to x$ for some $x \in A$. Since $\mathcal{U}$ is an (open) cover of A in X, we can pick some $j \in I$ such that $x \in U_j$. Since U_j is open, there exists $r > 0$ such that $B(x, r) \subseteq U_j$. By Archimedean property of $\mathbb{R}$, choose $n_o \in \mathbb{N}$ such that $\frac{1}{n} < \frac{r}{2} \ \forall \ n \geq n_o$. Since $x_{k_n} \to x$, there exists $m \geq n_o$ such that $d(x, x_{k_m}) < \frac{1}{k_m} \leq \frac{1}{m} < \frac{r}{2}$. It is easy to check that $B(x_{k_m}, \frac{1}{k_m}) \subseteq B(x, r)$. But $B(x, r) \subseteq U_j$. So $B(x_{k_m}, \frac{1}{k_m}) \cap (X - U_j) = \emptyset$ which is a contradiction to $(\star)$. Hence every open cover of A in (X, d) has a Lebesgue number. $\square$

Remark. Here note that the hypothesis is: every sequence in A has a subsequence converging to a point in A. This is stronger than A being totally bounded. Recall that A is totally bounded if and only if every sequence in A has a Cauchy subsequence.

We know that every bounded subset of $\mathbb{R}$ is totally bounded. So $(0, 1)$ is totally bounded in $\mathbb{R}$. *Find an open cover of $A = (0, 1)$ in $(\mathbb{R}, |\cdot|)$ with no Lebesgue number. What about $[0, 1]$? Can you prove that every open cover of $[0, 1]$ in $(\mathbb{R}, |\cdot|)$ has a Lebesgue number?*

Interestingly, now we are going to prove that the property (in terms of sequences) being satisfied by the subset A in Proposition 4.2.2 is equivalent to the compactness of A.

Theorem 4.2.1. *Let (X, d) be a metric space and A be an infinite subset of X. Then the following statements are equivalent:*

(a) *A is compact in (X, d).*
(b) *Every infinite subset of A has an accumulation point in A.*
(c) *Every sequence (x_n) in A has a subsequence (x_{k_n}) converging to a point in A.*

Proof. (a) $\Rightarrow$ (b): Let S be an infinite subset of the compact set A. If possible, suppose that S has no accumulation point in A. So if $x \in A$, then x is not an accumulation point of S. Hence there exists $r_x > 0$ such that $B(x, r_x) \cap (S \setminus \{x\}) = \emptyset$, that is, $B(x, r_x) \cap S \subseteq \{x\}$. Now $\{B(x, r_x) : x \in A\}$ is an open cover of the compact set A. Hence there exist $x_1, \ldots, x_n$ in A such that $A \subseteq \bigcup_{i=1}^{n} B(x_i, r_{x_i})$. Now $S = A \cap S \subseteq (\bigcup_{i=1}^{n} B(x_i, r_{x_i})) \cap S = \bigcup_{i=1}^{n} (B(x_i, r_{x_i}) \cap S) \subseteq \{x_1, \ldots, x_n\}$ which shows that S must be a finite set. This contradicts our assumption that S is an infinite subset of A. Hence every infinite subset of A has an accumulation point in A.

(b) $\Rightarrow$ (c): This is precisely Corollary 1.5.7.

(c) $\Rightarrow$ (a): Let $\{U_i : i \in I\}$ be an open cover of A in (X, d). Then by Proposition 4.2.2, $\{U_i : i \in I\}$ has a Lebesgue number $\delta > 0$. We claim that there exist $x_1, \ldots, x_n$ in A such that $A \subseteq \bigcup_{i=1}^{n} B(x_i, \delta)$. To see this, if possible, suppose that the claim is false. Fix some $x_1 \in A$ and then choose $x_2 \in A \setminus B(x_1, \delta)$. In general, using induction, choose $x_{n+1} \in A \setminus \bigcup_{i=1}^{n} B(x_i, \delta)$. Clearly, $d(x_n, x_m) \geq \delta$ for $n \neq m$. This implies that no subsequence of (x_n) can converge, which contradicts the hypothesis (c). Hence the claim holds. Now for each j, $1 \leq j \leq n$, pick some $i_j \in I$ such that $B(x_j, \delta) \subseteq U_{i_j}$. Then $A \subseteq \bigcup_{i=1}^{n} B(x_i, \delta) \subseteq \bigcup_{j=1}^{n} U_{i_j}$. This shows that A is compact in (X, d). $\qquad\square$

Consequently, Proposition 4.2.2 can be restated as: every open cover of a compact metric space has a Lebesgue number.

Corollary 4.2.2. *A metric space (X, d) is compact if and only if every countable open cover of (X, d) has a finite subcover.*

Proof. It is enough to prove the sufficient part. For that we use the sequential characterization of compactness given in Theorem 4.2.1. Suppose, if possible, there exists a sequence (x_n) of distinct terms in X with no convergent subsequence. Therefore by Theorem 1.6.1, (x_n) has no cluster point. This implies that for all $n \in \mathbb{N}$, there exists $r_n > 0$ such that $B(x_n, r_n) \cap S = \{x_n\}$, where $S = \{x_n : n \in \mathbb{N}\}$. Since (x_n) has no cluster point, S is closed. Now it is evident that $(X \setminus S) \bigcup \{B(x_n, r_n) : n \in \mathbb{N}\}$ is a countable open cover of X with no finite subcover. A contradiction! $\qquad\square$

Note that the metric spaces for which every countable open cover has a finite subcover are known as *countably compact*. Thus, for metric spaces the notion of compactness and that of its countable counterpart coincide. But here we would like to mention that this coincidence need not hold true in a general topological space.

In Proposition 4.2.1, it was seen that every compact metric space is totally bounded. But note that the converse is not true in general. For example, any open and bounded interval of $\mathbb{R}$ is totally bounded but not compact (it is not even closed). The next result highlights the gap between compact metric spaces and totally bounded metric spaces.

Theorem 4.2.2. *A metric space* (X, d) *is compact if and only if* (X, d) *is complete and totally bounded.*

Proof. Suppose (X, d) is compact and let (x_n) be a Cauchy sequence in (X, d). Then by Theorem 4.2.1, (x_n) has a convergent subsequence in X. Thus, the sequence (x_n) itself converges as it is Cauchy. Hence (X, d) is complete, whereas the total boundedness of (X, d) follows from Proposition 4.2.1.

Conversely, suppose that (X, d) is complete as well as totally bounded. We show that every sequence (x_n) in (X, d) has a convergent subsequence. Since (X, d) is totally bounded, (x_n) has a Cauchy subsequence (x_{k_n}) by Theorem 3.2.3. Again since (X, d) is complete, there exists $x \in X$ such that $x_{k_n} \to x$ in (X, d). Hence every sequence (x_n) in X has a convergent subsequence. So by Theorem 4.2.1, (X, d) is compact. $\qquad\square$

If (X, d) is a compact metric space, then the equivalent metrics will have to behave in a similar fashion. Suppose (X, d) is a compact metric space and ρ is a metric equivalent to d. Since the topology generated by ρ is same as the topology generated by d, (X, ρ) is also compact and consequently (X, ρ) is complete. But in case of a non-compact metric space, we may have two equivalent metrics d and ρ such that d is complete, while ρ is not complete. For example, on $(-\frac{\pi}{2}, \frac{\pi}{2})$, the usual distance metric $|\cdot|$ is not complete, while there exists a metric ρ, equivalent to $|\cdot|$, on $(-\frac{\pi}{2}, \frac{\pi}{2})$ such that ρ is complete (see Example 2.6.2). Note that $(-\frac{\pi}{2}, \frac{\pi}{2})$ with the usual distance $|\cdot|$ is not compact as it is not closed in $\mathbb{R}$. In the next section, it is proved that $[-\frac{\pi}{2}, \frac{\pi}{2}]$ with the usual distance is compact.

We would like to state a relevant result without proof. Interested readers may refer to [42] for the proof.

Theorem 4.2.3. *Let (X, d) be a metric space. Then (X, d) is compact if and only if for every equivalent metric ρ on X, (X, ρ) is complete.*

Exercises

4.2.1. Give an example of a complete metric space which is not compact.

4.2.2. Let E be a non-empty compact subset in (X, d). Let $\delta = \mathrm{diam}(E)$. Show that there exist x_o, $y_o \in E$ such that $d(x_o, y_o) = \delta$.

4.2.3. Show that the closed unit ball in $C[0, 1]$ with the supremum metric is not compact.

4.2.4. Let $\{(X_i, d_i) : 1 \le i \le n\}$ be a finite collection of metric spaces. Then show that the product space $\prod_{i=1}^{n} X_i$, equipped with any of the metrics given in Example 1.3.4, is compact if and only if (X_i, d_i) is compact for all $1 \le i \le n$.

4.2.5. Let $\{(X_i, d_i) : i \in \mathbb{N}\}$ be a countable collection of metric spaces. Then show that the Cartesian product $\prod_{i=1}^{\infty} X_i$, equipped with the metric given in Example 1.3.10, is compact if and only if (X_i, d_i) is compact for all $i \in \mathbb{N}$.

A generalized version of this result in topological space setting is well known as *Tychonoff Theorem*.

4.2.6. Prove that a metric space (X, d) is totally bounded if and only if its completion is compact.

4.2.7. A subset A of (X, d) is called *relatively compact* if $\overline{A}$ is compact. Now prove the following statements:

(1) Finite union of relatively compact subsets of (X, d) is also relatively compact.
(2) If A is relatively compact in (X, d) and $B \subseteq A$, then B is also relatively compact.
(3) A is relatively compact if and only if each sequence in A has a cluster point in X.

(4) Every relatively compact set is totally bounded. Give an example of a totally bounded set in some metric space that is not relatively compact.

4.3 Compactness *vis-à-vis* Continuity

Now think about the structure of the range of a continuous function defined on a closed and bounded interval of $\mathbb{R}$. We present the next two results in this regard, but the complete answer will follow once we talk about Heine–Borel Theorem.

Theorem 4.3.1. *Let* $f : (X,d) \to (Y,\sigma)$ *be a continuous map between two metric spaces. If* A *is compact in* (X,d), *then* $f(A)$ *is compact in* (Y,σ).

Proof. Let A be compact in (X,d). To show that $f(A)$ is compact in (Y,σ), let $\{V_i : i \in I\}$ be an open cover of $f(A)$ in (Y,σ), that is, $f(A) \subseteq \bigcup_{i \in I} V_i$ and V_i is open in (Y,σ) for each $i \in I$. Now

$$f(A) \subseteq \bigcup_{i \in I} V_i \Rightarrow A \subseteq f^{-1}\left(\bigcup_{i \in I} V_i\right) = \bigcup_{i \in I} f^{-1}(V_i).$$

Since f is continuous and each V_i is open in (Y,σ), $f^{-1}(V_i)$ is open in (X,d) $\forall\, i \in I$. Hence $\{f^{-1}(V_i) : i \in I\}$ is an open cover of A in (X,d). Since A is compact in (X,d), there exist $i_1, i_2, \ldots, i_k$ in I such that $A \subseteq \bigcup_{j=1}^{k} f^{-1}(V_{i_j})$. Hence,

$$f(A) \subseteq f\left(\bigcup_{j=1}^{k} f^{-1}(V_{i_j})\right) = \bigcup_{j=1}^{k} f(f^{-1}(V_{i_j})) \subseteq \bigcup_{j=1}^{k} V_{i_j}.$$

Hence $\{V_{i_j} : 1 \leq j \leq k\}$ is a finite subcover of $\{V_i : i \in I\}$ for $f(A)$. Thus, $f(A)$ is compact in (Y,σ). $\qquad\square$

Before stating the next result, recall that if A is a subset of $\mathbb{R}$ which is bounded above, then $\sup A$ is a real number and further, by definition of supremum, $\sup A \in \overline{A}$ in $(\mathbb{R}, |\cdot|)$. Similarly, if B is a subset of $\mathbb{R}$ which is bounded below, then $\inf B$ is a real number which lies in $\overline{B}$. In particular, if A is closed and bounded in $(\mathbb{R}, |\cdot|)$, then both $\sup A$ and $\inf A$ are in A itself. Now the next result is immediate from Theorem 4.3.1 and Proposition 4.1.2.

Theorem 4.3.2. *Let $f : (X, d) \to (\mathbb{R}, |\cdot|)$ be a real-valued continuous function. If A is compact in (X, d), then $f(A)$ is closed and bounded in $(\mathbb{R}, |\cdot|)$. Moreover, there exist a, $b \in A$ such that $\sup f(A) = f(a)$ and $\inf f(A) = f(b)$.*

In other words, it can be said that every real-valued continuous function, defined on a compact metric space, attains its extreme values. The reader should compare it with the "Extreme Value Theorem" of real analysis in which we particularly take a continuous function on a closed and bounded interval.

Theorem 4.3.3. *Let $f : (X, d) \to (Y, \sigma)$ be a continuous bijection between two metric spaces. If (X, d) is compact, then $f^{-1} : (Y, \sigma) \to (X, d)$ is also continuous and hence f is a homeomorphism.*

Proof. First let A be closed in (X, d). Since (X, d) is compact, by Proposition 4.1.4 A is compact in (X, d). Now Theorem 4.3.1 implies that $f(A)$ is compact in (Y, σ) and hence $f(A)$ is closed in (Y, σ). But $f(A) = (f^{-1})^{-1}(A)$. So if A is closed in (X, d), then $(f^{-1})^{-1}(A) = f(A)$ is closed in (Y, σ). Hence f^{-1} is continuous by Theorem 2.1.1. $\qquad\square$

Theorem 4.3.4. *Let (X, d) be a metric space. Then the following statements are equivalent:*

(a) *(X, d) is compact.*
(b) *(X, ρ) is bounded for every metric ρ equivalent to d.*
(c) *Every continuous function on (X, d) with values in an arbitrary metric space is bounded.*
(d) *Every real-valued continuous function on (X, d) is bounded.*

Proof. (a) $\Rightarrow$ (b): Let ρ be a metric on X which is equivalent to d. Then the identity map from (X, d) to (X, ρ) is a homeomorphism. Hence by Theorem 4.3.1, (X, ρ) is also compact. This implies that (X, ρ) is bounded by Proposition 4.2.1.

(b) $\Rightarrow$ (c): Let $f : (X, d) \to (Y, \sigma)$ be a continuous function. Now consider the function $\rho(x, y) = d(x.y) + \sigma(f(x), f(y))$ defined on $X \times X$. Then we know that (see Exercise 2.6.2) ρ is a metric on

X which is equivalent to d. By (b), (X, ρ) is bounded. Hence there exist $M > 0$ and $x_o \in X$ such that $\rho(x, x_o) < M$ for all $x \in X$. Consequently, $\sigma(f(x), f(x_o)) \leq \rho(x, x_o) < M$ for all $x \in X$. This implies that f is bounded.

(c) $\Rightarrow$ (d): This is immediate.

(d) $\Rightarrow$ (a): Suppose every real-valued continuous function on (X, d) is bounded. Let (x_n) be a sequence in X with no convergent subsequence. Then by Theorem 1.6.1, (x_n) has no cluster point. Without loss of generality, it can be assumed that $x_n \neq x_m \; \forall \, n \neq m$. Thus, the set $A = \{x_n : n \in \mathbb{N}\}$ is closed and the function $f : A \to \mathbb{R}$ defined as $f(x_n) = n$ is continuous. By Tietze's extension theorem, there exists a continuous function $F : X \to \mathbb{R}$ such that $F(a) = f(a) \; \forall \, a \in A$. Thus, by definition F is unbounded. This gives a contradiction to (d). Hence by Theorem 4.2.1, (X, d) is compact. $\qquad\square$

Now we would like to establish a link between Lebesgue numbers and compact sets in a metric space. Using the sequential characterization of compact sets, Proposition 4.2.2 can be reformulated in the following manner. But here we give a direct proof, just by using the definition of compactness.

Theorem 4.3.5. *Let (X, d) be a metric space and A be a non-empty compact subset of X. Then every open cover of A in (X, d) has a Lebesgue number.*

Proof. Let $\mathcal{U} = \{U_i : i \in I\}$ be an open cover of A in (X, d) and let $x \in A$. Since $\mathcal{U}$ is a cover of A and each U_i is open in (X, d), there exists some $U_x \in \mathcal{U}$ and $\epsilon_x > 0$ such that $x \in U_x$ and $B(x, 2\epsilon_x) \subseteq U_x$. Now $\{B(x, \epsilon_x) : x \in A\}$ is an open cover of the compact set A in (X, d). Hence there exists a finite set $\{x_1, x_2, \ldots, x_k\} \subseteq A$ such that $A \subseteq \bigcup_{i=1}^{k} B(x_i, \epsilon_{x_i})$. Now choose $\epsilon = \min_{1 \leq i \leq k} \epsilon_{x_i}$ and our claim is that ϵ is a Lebesgue number of the open cover $\mathcal{U}$ of A. Let $a \in A$. So there exists some $j \in \{1, 2, \ldots, k\}$ such that $a \in B(x_j, \epsilon_{x_j})$. We claim that $B(a, \epsilon) \subseteq B(x_j, 2\epsilon_{x_j})$. Take $y \in B(a, \epsilon)$. So $d(a, y) < \epsilon$. Now $d(x_j, y) \leq d(x_j, a) + d(a, y) < \epsilon_{x_j} + \epsilon \leq 2\epsilon_{x_j}$, that is, $y \in B(x_j, 2\epsilon_{x_j})$ and hence $B(a, \epsilon) \subseteq B(x_j, 2\epsilon_{x_j})$. But $B(x_j, 2\epsilon_{x_j}) \subseteq U_{x_j}$. Hence the claim is proved. $\qquad\square$

The converse of Theorem 4.3.5 is not true in general. In fact, in the next chapter we will have a detailed discussion on the metric spaces in which every open cover has a Lebesgue number.

Recall from real analysis that every continuous function on a closed and bounded interval $[a, b]$ is uniformly continuous. If we carefully analyze the proof then it can be observed that the compactness of $[a, b]$ is not required in full strength. Let us prove its generalized version.

Theorem 4.3.6. *Let $f : (X, d) \to (Y, \rho)$ be a continuous function. If every open cover of X has a Lebesgue number, then f is uniformly continuous.*

Proof. Let $\epsilon > 0$. By the continuity of f, for every $x \in X$, there exists $r_x > 0$ such that $d(x, y) < r_x \Rightarrow \rho(f(x), f(y)) < \frac{\epsilon}{2}$. Now $\{B(x, r_x) : x \in X\}$ is an open cover of X. Hence by the hypothesis, this open cover has a Lebesgue number, say $\delta > 0$.

Now assume that $x, x' \in X$ with $d(x, x') < \delta$. So $\exists z \in X$ such that $B(x', \delta) \subseteq B(z, r_z)$, which implies that $d(x', z) < r_z$. Also note that $x \in B(x', \delta)$. Hence $d(x, z) < r_z$. Recall how we got r_z. Hence $\rho(f(z), f(x')) < \frac{\epsilon}{2}$ and $\rho(f(z), f(x)) < \frac{\epsilon}{2}$. Consequently,

$$\rho(f(x), f(x')) \leq \rho(f(x), f(z)) + \rho(f(z), f(x')) < \frac{\epsilon}{2} + \frac{\epsilon}{2},$$

whenever $d(x, x') < \delta$. Hence f is uniformly continuous. $\qquad\square$

The next result is immediate from Theorems 4.3.5 and 4.3.6.

Theorem 4.3.7. *Let $f : (X, d) \to (Y, \rho)$ be a continuous function between two metric spaces. If (X, d) is compact, then f is uniformly continuous.*

But we would like to give a direct proof of the previous theorem, just by using the definition of compactness.

Alternative proof of Theorem 4.3.7. Let $\epsilon > 0$. Choose any $x \in X$. Since f is continuous at x, there exists $r_x > 0$ such that

$$d(x, x') < 2r_x, \ x' \in X \ \Rightarrow \ \rho(f(x), f(x')) < \epsilon/2. \quad (\star)$$

Now $\{B(x, r_x) : x \in X\}$ is an open cover of X. Since X is compact, there exist $x_1, \dots, x_n$ in X such that $X = \bigcup_{i=1}^n B(x_i, r_{x_i})$. Note that

X is also acting as the index set for the cover $\{B(x, r_x) : x \in X\}$. Let $\delta = \min_{1 \le i \le n} r_{x_i}$. Now let w, $w' \in X$ such that $d(w, w') < \delta$. So there exists x_i $(1 \le i \le n)$ such that $w \in B(x_i, r_{x_i})$, that is, $d(w, x_i) < r_{x_i}$. By $(\star)$, $\rho(f(x_i), f(w)) < \epsilon/2$ $(\star\star)$. Now

$$d(w', x_i) \le d(w', w) + d(w, x_i) < \delta + r_{x_i} \le 2r_{x_i}.$$

But $d(w', x_i) < 2r_{x_i} \Rightarrow \rho(f(x_i), f(w')) < \epsilon/2$ (by $(\star)$). Hence

$$\rho(f(w), f(w')) \le \rho(f(w), f(x_i)) + \rho(f(x_i), f(w'))$$
$$< \epsilon/2 + \epsilon/2 = \epsilon,$$

whenever $d(w, w') < \delta$.

Remark. In the beginning of this chapter, it has been said that compactness is a topological generalization of finiteness. Look at the alternative proof of Theorem 4.3.7 — check carefully where exactly the compactness of X is needed. In the absence of the hypothesis of compactness of X, we could have chosen δ to be $\inf_{x \in X} r_x$. But then δ could be 0, which won't work. In order to ensure that $\delta > 0$, we needed only a finite number of r_x. One can make the same observation in the proof of Theorem 4.3.5 as well where $\epsilon = \min_{1 \le i \le k} \epsilon_{x_i}$ was chosen.

Corollary 4.3.1. *Let d and ρ be two equivalent metrics on X such that (X, d) is compact. Then the two metrics are uniformly equivalent on X.*

Exercises

4.3.1. Suppose A is a subset of a compact metric space (X, d). Moreover, for every continuous function $f : (X, d) \to \mathbb{R}$, the restriction of f to A attains the maximum on A. Prove that A is compact.

4.3.2. Let (X, d) be a compact metric space and $f : (X, d) \to \mathbb{R}$ be a function which is upper semi-continuous at every point of X. Then show that f has an absolute maximum on X, that is, there exists $x_o \in X$ such that $f(x) \le f(x_o)$ for all $x \in X$.

4.3.3. Let $f : (X, d) \to (Y, \rho)$ be a function which is continuous on every compact subset of X. Show that f is continuous.

4.3.4. Let (X, d) be a compact metric space and $f : X \to X$ be an isometry. Then prove that f is onto. Does the conclusion hold if X is not compact?

4.3.5. (a) Let (X, d) be a compact metric space and $f : X \to X$ be a function such that $d(f(x), f(y)) < d(x, y)$ for $x \neq y$. Then show that f has a unique fixed point. Compare with Remark (iii) given after Theorem 2.4.2.

(b) Give an example of a compact metric space (X, d) and a function $f : X \to X$ such that $d(f(x), f(y)) \leq d(x, y)$ for all $x, \ y \in X$ and

   (i) f has no fixed point;
   (ii) f has more than one fixed point.

4.3.6. Can you find a homeomorphism between a parabola and a circle in $\mathbb{R}^2$?

4.3.7. Let A and B be two non-empty disjoint subsets of a metric space (X, d).

(a) If A and B are compact then show that $d(A, B) > 0$.
(b) Does the conclusion hold true if either of A or B is just closed?
(c) If A and B are closed, then show with the help of an example that $d(A, B)$ can be 0.

4.3.8. Give examples to show that compactness of the subset A in Theorem 4.3.2 cannot be dropped.

4.3.9. Let $f : (X, d) \to (Y, \rho)$ be a function. Then the graph G of f is a subset of $X \times Y$, defined by

$$G = \{(x, y) \in X \times Y : y = f(x)\}.$$

Suppose (Y, ρ) is compact, then show that f is continuous if and only if G is a closed subset of $X \times Y$, where $X \times Y$ is equipped with the metric:

$$D((x, y), (u, v)) = d(x, u) + \rho(y, v) \text{ for } (x, y), \ (u, v) \in X \times Y.$$

Give a counter example to show that if (Y, ρ) is not compact, then a function $f : (X, d) \to (Y, \rho)$ need not be continuous, even if its graph is closed in $(X \times Y, D)$.

4.3.10. Let I be a closed interval in $\mathbb{R}$. Prove that a function $f : I \to (Y, \rho)$ is continuous if and only if f is Cauchy-continuous.

4.3.11. Let f be a real-valued continuous function on a compact metric space (X, d) and $\epsilon > 0$. Then show that there exists $M > 0$ such that

$$|f(x) - f(y)| \le M d(x, y) + \epsilon \text{ for all } x,\ y \in X.$$

4.4 Heine–Borel Theorem

In Proposition 4.1.2, it was seen that every compact subset of a metric space is closed and bounded. In this section, we will talk about the converse. In general a closed and bounded subset of a metric space need not be compact. For example, $[0, \sqrt{3}] \cap \mathbb{Q}$ is closed and bounded in $\mathbb{Q}$ but not compact (use sequential characterization of compact spaces). Furthermore, in Exercise 4.1.4 it was seen that the closed unit ball is not compact in $(\ell^\infty(\mathbb{R}), D)$. Interestingly, in $\mathbb{R}$ every closed and bounded subset is compact as proved in the next theorem. In the literature, one can find many proofs for this theorem. But here an elementary and easy proof (just by using the Cantor's Intersection Theorem) is presented (see page 90 of [46]).

Theorem 4.4.1 (Heine–Borel theorem). *For $k \in \mathbb{N}$, a subset of the Euclidean space $\mathbb{R}^k$ is compact if and only if it is closed and bounded in $\mathbb{R}^k$.*

We defer the proof for a while.

Remarks. (a) If we consider the metric $d(x, y) = \min\{1, |x - y|\}$ on $\mathbb{R}$, then it can be easily verified that $(\mathbb{R}, d)$ is not compact but it is closed and bounded. This shows the significance of the standard metric in the Heine–Borel theorem, even uniformly equivalent metric does not work.

(b) Those metric spaces in which every closed and bounded subset is compact are referred to as **boundedly compact** metric spaces. Thus, by Heine–Borel theorem, the Euclidean space $\mathbb{R}^k$ is boundedly compact.

Recall that in any metric space, every compact subset is closed as well as bounded. Hence the strength of Heine–Borel theorem lies in the reverse direction, that is, any closed and bounded subset of $\mathbb{R}^k$ is compact in $\mathbb{R}^k$.

We first look at a very intuitive corollary of Heine–Borel theorem. In this regard, recall that there is an onto isometry from $\mathbb{C}^k$ to $\mathbb{R}^{2k}$ given by

$$f : \mathbb{C}^k \to \mathbb{R}^{2k}$$
$$z = (z_1, \ldots, z_k) \to f(z) = (x_1, y_1, x_2, y_2, \ldots, x_k, y_k)$$

where $z_i = x_i + \iota y_i$, $1 \leq i \leq k$ and both $\mathbb{C}^k$ and $\mathbb{R}^{2k}$ are equipped with usual distance metric. Subsequently, we have the following result.

Corollary 4.4.1. *A subset of $\mathbb{C}^k$ is compact if and only if it is closed and bounded in $\mathbb{C}^k$.*

For an easy proof of Heine–Borel theorem, we need to talk about cells in $\mathbb{R}^k$.

Definition 4.4.1. A non-empty subset F of $\mathbb{R}^k$ ($k \in \mathbb{N}$) is called a **cell** (more precisely a k-cell) if there exist closed intervals $[a_1, b_1]$, $[a_2, b_2], \ldots, [a_k, b_k]$ such that $F = [a_1, b_1] \times [a_2, b_2] \times \cdots \times [a_k, b_k]$, that is, $F = \{(x_1, \ldots, x_k) \in \mathbb{R}^k : x_j \in [a_j, b_j]$ for $1 \leq j \leq k\}$. Note that the diameter of F is $\delta = \sup\{d(x, y) : x, y \in F\} = \left(\sum_{j=1}^{k}(b_j - a_j)^2\right)^{1/2}$, where d is the Euclidean metric on $\mathbb{R}^k$.

Using midpoints $c_j = \frac{1}{2}(a_j + b_j)$ of $[a_j, b_j]$, we can see that F is a union of 2^k k-cells each having diameter $\delta/2$. For a better understanding, first consider the case of $k = 2$ (see Figure 4.1) and that of $k = 3$, that is, consider the usual plane and the three-dimensional Euclidean space.

Theorem 4.4.2. *Every k-cell in the Euclidean space $\mathbb{R}^k$ is compact.*

Proof. Let F be a cell in $\mathbb{R}^k$ and if possible, assume that F is not compact. Then there exists an open cover $\mathcal{U} = \{U_i : i \in I\}$, U_i is open in $\mathbb{R}^k$, of F such that no finite subfamily of $\mathcal{U}$ covers F. Let δ denotes the diameter of F. As noted earlier, F is a union of 2^k k-cells, each having diameter $\delta/2$. At least one of these 2^k k-cells, which we denote by F_1, cannot be covered by finitely many sets from $\mathcal{U}$. Likewise, F_1 contains a k-cell F_2 of diameter $\delta/4 = \delta/2^2$, which cannot be covered

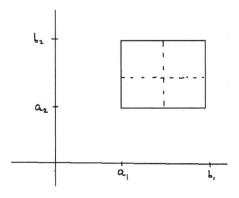

Figure 4.1 2-cell.

by finitely many sets from $\mathcal{U}$. Continuing in this fashion, we obtain a sequence (F_n) of k-cells such that

(i) $F \supseteq F_1 \supseteq F_2 \supseteq F_3 \supseteq \ldots$
(ii) F_n has diameter $\delta/2^n$
(iii) F_n cannot be covered by finitely many sets from $\mathcal{U}$.

Since $\mathbb{R}^k$ is a complete metric space, by Cantor's Intersection Theorem, $\bigcap_{n=1}^{\infty} F_n$ contains a point x_o. Recall that $\mathcal{U}$ is an (open) cover of F and $F_n \subseteq F$ $\forall\, n \in \mathbb{N}$. Hence $x_o \in F$ and consequently, there exists $U_o \in \mathcal{U}$ such that $x_o \in U_o$. Since U_o is open in $\mathbb{R}^k$, there exists an open ball $B(x_o, r)$ in $\mathbb{R}^k$ such that $x_o \in B(x_o, r) \subseteq U_o$. Now if $\delta/2^n < r$, one can easily see that $F_n \subseteq B(x_o, r)$ (because $x_o \in F_n$), that is, $F_n \subseteq U_o$ if $\delta/2^n < r$. But this contradicts (iii). Hence every k-cell in $\mathbb{R}^k$ is compact. $\qquad\square$

Now let us prove the main result.

Proof of Heine–Borel theorem. Let E be a closed and bounded set in $\mathbb{R}^k$. If E is empty, then E is trivially compact in $\mathbb{R}^k$. So, assume that E is non-empty. Since E is bounded, there exists a positive real number m such that E is contained in $\prod_{i=1}^{k}[-m, m]$, that is, in the set $F = \{(x_1, \ldots, x_k) \in \mathbb{R}^k : |x_j| \le m \text{ for } 1 \le j \le k\}$. By Theorem 4.4.2, F is compact in $\mathbb{R}^k$. Hence by Proposition 4.1.4, E is compact in $\mathbb{R}^k$.

Remark. For proving Heine–Borel theorem, the sequential characterization of a compact set in a metric space can also be used.

For the details, interested readers may refer to Theorem 7.4 (page 51) of [1].

Let us look at one of the nice applications of Heine–Borel theorem.

Proposition 4.4.1. *Let A be an uncountable subset of the Euclidean space $(\mathbb{R}^n, d)$. Then there exist infinitely many accumulation points of A in $\mathbb{R}^n$.*

Proof. First, we prove the existence of an accumulation point of A. Note that $A = \bigcup_{k \in \mathbb{N}} (A \cap C(\mathbf{0}, k))$, where $C(\mathbf{0}, k))$ denotes the closed ball in $\mathbb{R}^n$ around $\mathbf{0}$ with radius k. Since A is uncountable, $A \cap C(\mathbf{0}, k)$ is infinite for some $k \in \mathbb{N}$. By Heine–Borel theorem, $C(\mathbf{0}, k)$ is compact. Then by Theorem 4.2.1, $A \cap C(\mathbf{0}, k)$ (and hence A) has an accumulation point in $\mathbb{R}^n$.

Now we prove that there are infinitely many accumulation points of A. Suppose, if possible, there are only finitely many accumulation points of A in $\mathbb{R}^n$, namely $x_1, \ldots, x_k$. For $j \in \mathbb{N}$, let $S_j = A \cap \{x \in \mathbb{R}^n : d(x, x_i) > \frac{1}{j} \ \forall \ i \in \{1, 2, \ldots, k\}\}$. Consequently,

$$A \subseteq \bigcup_{j \in \mathbb{N}} S_j \cup \{x_1, \ldots, x_k\}.$$

This implies that S_j is uncountable for some $j \in \mathbb{N}$, since A is uncountable. Thus, by our previous claim, there exists an accumulation point of S_j, say $s \in \mathbb{R}^n$. But $\overline{S_j} \subseteq S_m$ for some m with $0 < j < m$. Then s is an accumulation point of A other than $x_1, \ldots, x_k$. A contradiction! $\qquad\square$

Historical Perspective

(i) Heinrich Eduard Heine (1821–1881) was a German mathematician. He worked on PDE and special functions. Heine introduced the concept of uniform continuity.

(ii) Emile Borel (1871–1956) was a French mathematician. He was one of the founders of the modern theory of functions along with Baire and Lebesgue. In fact, the measure theory was founded by these three mathematicians and Greek mathematician Caratheodory. It was in 1872 that Heine gave a proof of Heine–Borel Theorem, while Borel published his proof much later in 1895.

Exercises

Note. *The reader should keep in mind that unless mentioned explicitly, $\mathbb{R}^k$ is assumed to be equipped with the Euclidean metric.*

4.4.1. For $m \in \mathbb{N}$, let $A_m = \{(x,y) \in \mathbb{R}^2 : x^m + y^m = 1\}$. Use Heine–Borel Theorem to show that A_m is compact in $\mathbb{R}^2$ if and only if m is even.

4.4.2. Let A and B be two non-empty closed sets in $\mathbb{R}^k$.

(a) Give an example to show that $A + B = \{a + b : a \in A, b \in B\}$ may not be closed in $\mathbb{R}^k$.
(b) If A is compact in $\mathbb{R}^k$, then show that $A + B$ is closed in $\mathbb{R}^k$. But $A + B$ need not be compact in $\mathbb{R}^k$. Find a counter example.
(c) If A and B are both compact in $\mathbb{R}^k$, then prove that $A + B$ is also compact in $\mathbb{R}^k$.

4.4.3. Show that each boundedly compact metric space can be expressed as a union of an increasing sequence of compact subsets.

4.4.4. Show that each boundedly compact metric space is both separable and complete.

4.4.5. Let A be a closed subset of a boundedly compact metric space (X, d). Prove that (A, d) is boundedly compact.

4.4.6. Does the uniformly continuous image of a boundedly compact metric space also boundedly compact?

Chapter 5

Weaker Notions of Compactness

It is well known that compactness is a very strong property which is not possessed by a large collection of metric spaces. Naturally, in such cases our next step should be to look for some weaker property which could be useful in the study. In this chapter, our main goal is to make our readers familiar with four very interesting concepts: local compactness, UCness, cofinal completeness and finite chainability. All of them are weaker than compactness. But UC spaces and cofinally complete metric spaces are stronger than complete metric spaces, while finite chainability is stronger than boundedness.

5.1 Local Compactness

In Heine–Borel Theorem, it was seen that although $\mathbb{R}$ is not compact but for every $x \in X$ and $\epsilon > 0$, the interval $[x - \epsilon, x + \epsilon]$ is compact. Now if we look at $\mathbb{Q}$, then even for a single $x \in \mathbb{Q}$, we won't be able to find any $\epsilon > 0$ such that $[x - \epsilon, x + \epsilon] \cap \mathbb{Q}$ becomes compact. This gives an idea of a notion, weaker than compactness, which is applied in a *local* sense.

Definition 5.1.1. A metric space (X, d) is said to be **locally compact** if for every $x \in X$ there exists an $\epsilon > 0$ such that $\overline{B(x, \epsilon)}$ is compact.

Remark. Note that a subset A of a metric space (X, d) is said to be locally compact if the metric space (A, d) is locally compact.

Since a closed subset of a compact metric space is compact, every compact space is locally compact. But the converse is not true in general. For example, the Euclidean space $\mathbb{R}^n$ is locally compact (by Heine–Borel Theorem) but not compact. The next result says that we can call a metric space to be compact in a *local* sense if every $x \in X$ has 'arbitrarily small' compact neighbourhoods.

Theorem 5.1.1. *Let (X, d) be a metric space. Then (X, d) is locally compact if and only if for every $x \in X$ and $\epsilon > 0$, there exists a $\delta > 0$ such that $\overline{B(x, \delta)} \subseteq B(x, \epsilon)$ and $\overline{B(x, \delta)}$ is compact.*

Proof. Let (X, d) be locally compact. If $x \in X$ and $\epsilon > 0$, then there exists an $\epsilon_1 > 0$ such that $\overline{B(x, \epsilon_1)}$ is compact. Now we can choose a positive $\delta < \min\{\epsilon, \epsilon_1\}$ such that $\overline{B(x, \delta)} \subseteq B(x, \epsilon) \cap B(x, \epsilon_1)$. Since closed subset of a compact space is compact, $\overline{B(x, \delta)}$ is compact. Hence we are done.

The reverse implication follows immediately from the definition. $\square$

Now recall that $\mathbb{R}$ is locally compact, while its subset $\mathbb{Q}$ is not locally compact with the Euclidean distance (as per the discussion before Definition 5.1.1). Let us look at some additional conditions under which local compactness is preserved by the subsets.

Corollary 5.1.1. *An open or closed subset of a locally compact metric space is also locally compact.*

Proof. Let A be an open subset of a locally compact space (X, d). Suppose $a \in A$. Since A is open, there exists an $\epsilon > 0$ such that $B(a, \epsilon) \subseteq A$. Then by Theorem 5.1.1, there exists a $\delta > 0$ such that $\text{cl}_X(B(a, \delta)) \subseteq B(a, \epsilon)$ and $\text{cl}_X(B(a, \delta))$ is compact. Since $\text{cl}_X(B(a, \delta)) \subseteq B(a, \epsilon) \subseteq A$, $\text{cl}_A(B(a, \delta) \cap A) = \text{cl}_X(B(a, \delta))$. Consequently, (A, d) is locally compact.

Now assume that E is closed in (X, d) and let $e \in E$. We need to find some $\delta' > 0$ such that $\text{cl}_E(B(e, \delta') \cap E)$ is compact. By the local compactness of (X, d), there exists an $\epsilon' > 0$ such that $\text{cl}_X(B(e, \epsilon'))$ is compact. Note that $\text{cl}_E(B(e, \epsilon') \cap E) = \{\text{cl}_X(B(e, \epsilon') \cap E)\} \bigcap E$ is a closed subset of the compact space $\text{cl}_X(B(e, \epsilon'))$. Hence $\text{cl}_E(B(e, \epsilon') \cap E)$ is compact. $\square$

Thus, every open interval and closed interval in the Euclidean space $\mathbb{R}^n$ is also locally compact.

Proposition 5.1.1. *Let f be a continuous function from (X, d) onto (Y, ρ). If (X, d) is locally compact then so is (Y, ρ) provided f is an open map.*

Proof. Let $x \in X$. Then there exists an $\epsilon > 0$ such that $\overline{B_d(x, \epsilon)}$ is compact. Since f is continuous, Theorem 4.3.1 implies that $f(\overline{B_d(x, \epsilon)})$ is also compact. Now f being an open map, $f(B_d(x, \epsilon))$ is open. Since $f(x) \in f(B_d(x, \epsilon))$, there exists an $\epsilon_o > 0$ such that $B_\rho(f(x), \epsilon_o) \subseteq f(B_d(x, \epsilon)) \subseteq f(\overline{B_d(x, \epsilon)})$. The compactness of $f(\overline{B_d(x, \epsilon)})$ implies that it is closed and hence we have, $\overline{B_\rho(f(x), \epsilon_o)} \subseteq f(\overline{B_d(x, \epsilon)})$. This implies that $\overline{B_\rho(f(x), \epsilon_o)}$ is compact being a closed subset of a compact set. Since $Y = f(X)$, (Y, ρ) is locally compact. $\qquad\square$

Consequently, local compactness is a topological property, that is, it is preserved under homeomorphism.

Exercises

5.1.1. Show that each boundedly compact metric space is locally compact. Give an example of a metric subspace of the real line that is locally compact but not boundedly compact.

5.1.2. Verify that a metric space in which every point is an isolated point is locally compact.

5.1.3. Give an example to show that the condition of openness of the map f in Proposition 5.1.1 cannot be dropped for the conclusion to hold true.

5.1.4. Let $\{(X_i, d_i) : 1 \le i \le n\}$ be a finite collection of metric spaces. Then show that the product space $\prod_{i=1}^n X_i$, equipped with any of the metrics given in Example 1.3.4, is locally compact if and only if (X_i, d_i) is locally compact for all $1 \le i \le n$.

5.1.5. Let K be a compact subset of a locally compact metric space (X, d). If U is an open set containing K, then show that there exists an open set A such that $K \subseteq A \subseteq \overline{A} \subseteq U$ and $\overline{A}$ is compact.

5.1.6. (a) Give an example of a locally compact metric space which is not complete.

(b) Prove that every locally compact metric space is a Baire space. *This is Baire Category Theorem for locally compact spaces.*

(c) Give an example of a metric space which is a Baire space but not locally compact.

Note that using (a) and (b) we can construct examples of Baire spaces which are not complete.

5.2 UC Spaces

The significant role played by continuous function and its stronger version, namely uniformly continuous function, in the theory of analysis is unquestioned. This prompted many of the mathematicians to study the gap between the two classes of functions. Every uniformly continuous function is continuous, so the main question is: *under what equivalent conditions on a metric space (X, d), does every real-valued continuous function defined on (X, d) is uniformly continuous?*

It was proved that every continuous function from a compact metric space to an arbitrary metric space is uniformly continuous, but compactness is clearly not a necessary condition (consider any infinite set equipped with the discrete metric). In fact, it is a characteristic property of a larger class of metric spaces widely known as UC spaces or Atsuji spaces. In this section, some of the very interesting characterizations of UC spaces are presented. But first let us define them precisely.

Definition 5.2.1. A metric space (X, d) is called a **UC space** or an **Atsuji space** if every real-valued continuous function on (X, d) is uniformly continuous.

Thus by Theorem 4.3.7, every compact metric space is a UC space. In fact, more precisely the class of UC spaces lies between the class of compact metric spaces and that of complete metric spaces as proved in the next result. For that, we need to define uniformly discrete sets in a metric space. Recall the definition of a discrete set in a metric space. In Definition 1.3.2, if δ does not depend on x then we call it 'uniformly' discrete set. More precisely, we have

Definition 5.2.2. A subset A of a metric space (X, d) is called **uniformly discrete** if there exists a $\delta > 0$ such that $d(x, y) \geq \delta \, \forall \, x, \, y \in A, \, x \neq y$.

Proposition 5.2.1. *Every UC space is complete.*

Proof. Suppose (X, d) is UC but not complete. Then there exists a Cauchy sequence (x_n) of distinct points in X such that it does not converge in X. Thus, (x_n) has no cluster point by Theorem 1.6.1. Consequently, the set $A = \{x_n : n \in \mathbb{N}\}$ is closed and discrete in X.

Now define a function $f : A \to \mathbb{R}$ as follows: $f(x_n) = n \, \forall \, n \in \mathbb{N}$. Since A is discrete, f is continuous. By Tietze's extension theorem, f can be extended to a continuous function F on X. But F is not uniformly continuous because (x_n) is Cauchy in (X, d) and $(f(x_n)) = (n)$ is not Cauchy in $\mathbb{R}$ (see Proposition 2.3.1). A contradiction! $\square$

Note that there exists a complete metric space which is not UC (consider $\mathbb{R}$ with the usual distance metric). In Theorem 4.3.6, one of the sufficient conditions for a metric space to be UC is already proved: if every open cover of (X, d) has a Lebesgue number, then every continuous function $f : (X, d) \to (Y, \rho)$ is uniformly continuous. Interestingly, it can be shown that the converse of this result is also true.

Theorem 5.2.1. *For a metric space (X, d), the following conditions are equivalent:*

(a) (X, d) *is a UC space.*
(b) *If A_1 and A_2 are two disjoint nonempty closed sets in X, then $d(A_1, A_2) > 0$.*
(c) *Every closed discrete subset of X is uniformly discrete in X.*
(d) *Every open cover of X has a Lebesgue number.*

Proof. (a) $\Rightarrow$ (b): Let A_1 and A_2 be two disjoint nonempty closed sets in X with $d(A_1, A_2) = \inf\{d(x, y) : x \in A_1, \, y \in A_2\} = 0$. Then by definition of infimum, for all $n \in \mathbb{N}$, there exist $a_n \in A_1$ and $w_n \in A_2$ such that $d(a_n, w_n) < \frac{1}{n}$. By Urysohn's Lemma (Theorem 2.2.1), there exists a continuous function $f : X \to [0, 1]$ such that $f(a) = 0 \, \forall \, a \in A_1$ and $f(w) = 1 \, \forall \, w \in A_2$. But then f is not uniformly continuous which is a contradiction to (a).

(b) $\Rightarrow$ (c): Let A be a closed discrete subset of X which is not uniformly discrete. Thus for each n, there exist a_n, b_n in A, $a_n \neq b_n$, such that $d(a_n, b_n) < \frac{1}{n}$. Moreover we can assume that a_n and b_n are distinct from $2n - 2$ preceding points $a_1, \ldots, a_{n-1}, b_1, \ldots, b_{n-1}$. Otherwise, for each $m \geq n$ we get: whenever $0 < d(a, b) < \frac{1}{m}$, $a, b \in A$, we have either $a \in \{a_1, \ldots, a_{n-1}, b_1, \ldots, b_{n-1}\}$ or $b \in \{a_1, \ldots, a_{n-1}, b_1, \ldots, b_{n-1}\}$. Since $\{a_1, \ldots, a_{n-1}, b_1, \ldots, b_{n-1}\}$ is a finite set, we can find some $z \in \{a_1, \ldots, a_{n-1}, b_1, \ldots, b_{n-1}\}$ and a strictly increasing sequence (n_k) in $\mathbb{N}$ such that $\forall\, k \in \mathbb{N}$, there exists some $z_k \in A$ such that $0 < d(z_k, z) < \frac{1}{n_k}$. This implies that z is an accumulation point of A, which is a contradiction as A is discrete and $z \in A$. Hence, the sets $A_1 = \{a_n : n \in \mathbb{N}\}$ and $A_2 = \{b_n : n \in \mathbb{N}\}$ are disjoint. Also, A_1 and A_2 are closed in X as they are subsets of a closed and discrete set A. But $d(A_1, A_2) = 0$, which is a contradiction to (b).

(c) $\Rightarrow$ (d): Let $\mathcal{U} = \{U_i : i \in I\}$ be an open cover of X with no Lebesgue number. Thus, for all $n \in \mathbb{N}$, there exists $x_n \in X$ such that $B(x_n, \frac{1}{n}) \not\subseteq U_i \,\forall\, i \in I$. Since $\mathcal{U}$ is an open cover of X, each $x_n \in U_{i_n}$ for some $i_n \in I$ and there exists $y_n \in X$ with $d(x_n, y_n) < \frac{1}{n}$ but $y_n \notin U_{i_n}$. Since all U_i's are open, by passing to a subsequence we can assume that $x_n \neq x_m$ for $n \neq m$. Moreover, the set $A = \{x_n, y_n : n \in \mathbb{N}\}$ has no accumulation point: suppose if possible, z is an accumulation point of A. Then $B(z, \delta) \subseteq U_i$ for some $i \in I$ and some $\delta > 0$ (U_i is open). Also, there exists a subsequence of (x_n) which converges to z. Then $B(x_k, \frac{1}{k}) \subseteq B(z, \delta) \subseteq U_i$ for some $k \in \mathbb{N}$, which is a contradiction. Thus A has no accumulation point and hence it is a closed and discrete subset of X. By (c), A is uniformly discrete. Again a contradiction!

(d) $\Rightarrow$ (a): It follows from Theorem 4.3.6. $\qquad\square$

Remarks. (i) Conditions (b), (c) and (d) have been studied in [40], [44] and [57].

(ii) Due to the characterizing property (d), UC spaces are also known as *Lebesgue spaces*.

Here the reader should observe that the previous theorem is very helpful in discarding the *UCness* of a certain metric space.

For example, $[0, \infty)$ is not UC as $A = \{n, n + \frac{1}{n} : n \in \mathbb{N}\}$ is closed and discrete in $[0, \infty)$ but not uniformly discrete. Like we have completeness being characterized by Cauchy sequences, next we present some of the interesting sequential characterizations of UC spaces. First we give some required definitions.

Definition 5.2.3. A sequence (x_n) in a metric space (X, d) is called **pseudo-Cauchy** if it satisfies the following condition: $\forall \ \epsilon > 0$ and $\forall \ n \in \mathbb{N}$, there exist $j, k \in \mathbb{N}$ such that $j \neq k$, $j, k > n$ and $d(x_j, x_k) < \epsilon$.

In Cauchy sequences, the sequential terms gets arbitrary close eventually. Now if we analyze the definition of pseudo-Cauchy sequences, then in other words it can be said that frequently the sequential terms gets arbitrarily close. One of the very useful characterizations of UC spaces is in terms of the *isolation functional* $I(\cdot)$ defined by: $I(x) = d(x, X \setminus \{x\}) = \inf\{d(x, y) : y \in X \setminus \{x\}\}$. Thus, if x is an isolated point in X, then $I(x) = \sup\{r > 0 : B(x, r) = \{x\}\}$. Clearly, $I(x) = 0$ if and only if $x \in X'$.

Theorem 5.2.2. *For a metric space (X, d), the following conditions are equivalent:*

(a) *If (x_n) and (y_n) are two asymptotic sequences in X such that $x_n \neq y_n$ for each n, then the sequence (x_n) (equivalently (y_n)) has a cluster point in X.*

(b) *Every sequence (x_n) in X with $\lim\limits_{n \to \infty} I(x_n) = 0$ has a cluster point.*

(c) *Every pseudo-Cauchy sequence with distinct terms in X has a cluster point.*

(d) *(X, d) is a UC space.*

Proof. (a) $\Rightarrow$ (b): Let (x_n) be a sequence with $\lim\limits_{n \to \infty} I(x_n) = 0$. By passing to a subsequence, we can assume that $I(x_n) < \frac{1}{n}$. Thus, for all $n \in \mathbb{N}$, there exists $y_n \in X$ such that $0 < d(x_n, y_n) < \frac{1}{n}$. Then, (x_n) and (y_n) are asymptotic sequences and hence (x_n) clusters in X.

(b) $\Rightarrow$ (c): Let (x_n) be a pseudo-Cauchy sequence with distinct terms in X. Then we can find a subsequence (x_{k_n}) of (x_n) such that

$I(x_{k_n}) \to 0$. Consequently, the sequence (x_{k_n}), and hence (x_n), has a cluster point.

(c) $\Rightarrow$ (d): Let $f : (X, d) \to \mathbb{R}$ be a continuous function which is not uniformly continuous. Then there exists an $\epsilon_o > 0$ such that for all $n \in \mathbb{N}$, there exist $x_n, y_n \in X$ with $0 < d(x_n, y_n) < \frac{1}{n}$ but $|f(x_n) - f(y_n)| > \epsilon_o$ ($\star$). We can assume that $x_n \neq x_m$, $y_n \neq y_m$ and $x_n \neq y_m$ for $n \neq m$, otherwise we will get contradiction to the continuity of f (like we did in the proof of (b) $\Rightarrow$ (c) of Theorem 5.2.1). Thus, the sequence (z_n), where $z_{2n-1} = x_n$ and $z_{2n} = y_n$ for all $n \in \mathbb{N}$, is pseudo-Cauchy with distinct terms and hence it has a cluster point, say x_o. Since f is continuous at x_o, there exists a $\delta > 0$ such that $|f(x) - f(x_o)| < \frac{\epsilon_o}{2}$ whenever $x \in B(x_o, \delta)$. Then for sufficiently large n, $x_n, y_n \in B(x_o, \delta)$, which implies that $|f(x_n) - f(y_n)| \leq |f(x_n) - f(x_o)| + |f(y_n) - f(x_o)| < \epsilon_o$. This gives a contradiction to ($\star$).

(d) $\Rightarrow$ (a): Suppose there exist two asymptotic sequences (x_n) and (y_n) in X such that $x_n \neq y_n$ for each n and (x_n) (equivalently (y_n)) has no cluster point. Hence by passing to a subsequence, we can assume that $x_n \neq x_m$ and $y_n \neq y_m$ for $n \neq m$. Now we construct two subsequences (x'_{n_k}) and (y'_{n_k}) such that $x'_{n_k} \neq y'_{n_l} \; \forall \, k, l \in \mathbb{N}$ and $d(x'_{n_k}, y'_{n_k}) < 1/k$.

We proceed by induction. For $n = 1$, let $x'_{n_1} = x_1$ and $y'_{n_1} = y_1$. Let $A_1 = \{x'_{n_1}\}$ and $B_1 = \{y'_{n_1}\}$. Then, $A_1 \cap B_1 = \emptyset$. Suppose that we have chosen $x'_{n_1}, \ldots, x'_{n_k}$ and $y'_{n_1}, \ldots, y'_{n_k}$ and $A_k = \{x'_{n_1}, \ldots, x'_{n_k}\}$, $B_k = \{y'_{n_1}, \ldots, y'_{n_k}\}$ are such that $A_k \cap B_k = \emptyset$. Then choose $n_{k+1} = \min\{m > n_k : x_m \notin A_k \cup B_k \text{ or } y_m \notin A_k \cup B_k\}$.

Suppose that n_{k+1} does not exist, that is, $\forall \, n > n_k$, $x_n, y_n \in A_k \cup B_k$. This implies that $\{d(x_n, y_n) : n > n_k\}$ is finite and hence $\lim_{n \to \infty} d(x_n, y_n) > 0$ which contradicts our choice of (x_n) and (y_n).

If $x_{n_{k+1}} \in B_k$ or $y_{n_{k+1}} \in A_k$, then let $x'_{n_{k+1}} = y_{n_{k+1}}$ and $y'_{n_{k+1}} = x_{n_{k+1}}$, otherwise let $x'_{n_{k+1}} = x_{n_{k+1}}$ and $y'_{n_{k+1}} = y_{n_{k+1}}$.

Let $A = \{x'_{n_k} : k \in \mathbb{N}\}$ and $B = \{y'_{n_k} : k \in \mathbb{N}\}$. Since the sets A and B have no accumulation point in X, A and B are closed in X. Also by construction of (x'_{n_k}) and (y'_{n_k}), $A \cap B = \emptyset$ and $d(A, B) = 0$. By Urysohn's lemma (Theorem 2.2.1), we can have a real valued continuous function, $f : X \longrightarrow [0, 1]$ such that $f(A) = \{0\}$ and

$f(B) = \{1\}$. By (d), f is uniformly continuous, which is a contradiction as $d(A, B) = 0$. $\square$

Remarks. (i) Conditions (a), (b) and (c) have been mentioned in [57], [30] and [56] respectively.
(ii) In Theorem 5.2.2(c), it is important to take pseudo-Cauchy sequence of *distinct terms*. For example, consider the UC space $(\mathbb{N}, | \cdot |)$. Here observe that the sequence $\langle 1, 1, 2, 2, 3, 3, \ldots \rangle$ is pseudo-Cauchy with no cluster point.

Corollary 5.2.1. *Every closed subset of a UC space is UC.*

Proof. Let A be a closed subset of a UC space (X, d) and let (x_n) be a pseudo-Cauchy sequence of distinct terms in A. Since (X, d) is a UC space, (x_n) has a cluster point in X. But since A is closed in X, this cluster point must belong to A. Hence by Theorem 5.2.2, (A, d) is a UC space. $\square$

It should be noted that for defining UC spaces we do not need to restrict the range space to $\mathbb{R}$ as proved in the following result.

Corollary 5.2.2. *For a metric space (X, d), the following statements are equivalent:*

(a) (X, d) *is a UC space.*
(b) *Given any metric space (Y, ρ), any continuous function $f :$ $(X, d) \to (Y, \rho)$ is uniformly continuous.*

Proof. We only need to prove $(a) \Rightarrow (b)$. Let $f : (X, d) \to (Y, \rho)$ be any continuous function which is not uniformly continuous. Then there exists an $\epsilon_o > 0$ such that for all $n \in \mathbb{N}$, $\exists\, x_n, w_n \in X$ with $d(x_n, w_n) < \frac{1}{n}$ but $\rho(f(x_n), f(w_n)) > \epsilon_o$. Then $I(x_n) \to 0$ and hence by Theorem 5.2.2, (x_n) has a cluster point, say x_o. Since f is continuous at x_o, there exists a $\delta > 0$ such that $\rho(f(x), f(x_o)) < \frac{\epsilon_o}{2}$ whenever $x \in B(x_o, \delta)$. Now for sufficiently large n, $x_n, w_n \in B(x_o, \delta)$. Consequently, $\rho(f(x_n), f(w_n)) \leq \rho(f(x_n), f(x_o)) + \rho(f(w_n), f(x_o)) < \epsilon_o$. We arrive at a contradiction. $\square$

Interested reader can refer to the survey article [34] by Kundu and Jain for a well-organized collection of 25 equivalent characterizations of a UC space.

Corollary 5.2.3. *Suppose (X, d) is a UC space. Then the set X' of all accumulation points of X is compact.*

Proof. If X' is finite, then we are done. Suppose X' is infinite. We use Theorem 4.2.4 to prove the compactness of X'. Let (x_n) be a sequence in X'. Then $I(x_n) = 0$ for all $n \in \mathbb{N}$. Since (X, d) is UC, (x_n) has a cluster point in X by Theorem 5.2.2. Now Theorem 1.6.1 implies that there exists a subsequence (x_{k_n}) of (x_n) such that $x_{k_n} \to x$ for some $x \in X$. Since $(x_{k_n}) \subseteq X'$, $x \in X'$. Thus by Theorem 4.2.4, X' is compact. $\qquad \square$

Remark. It should be noted that the converse of Corollary 5.2.3 is not true in general. For example, consider $X = [0, 1] \cup A$, where $A = \{n, \ n + \frac{1}{n} : n \in \mathbb{N}\}$, equipped with the usual distance metric. Then $X' = [0, 1]$ which is compact. But $(X, |\cdot|)$ is not UC. *Find a pseudo-Cauchy sequence of distinct points in X which does not cluster?*

Although the compactness of the set of accumulation points X' does not imply that the metric space (X, d) is UC, but at least it guarantees the existence of an equivalent metric ρ on X such that (X, ρ) is UC.

Theorem 5.2.3. *Let (X, d) be a metric space such that X' is compact. Then there exists an equivalent metric ρ on X such that (X, ρ) is a UC space.*

Proof. If $X' = \emptyset$, then take ρ to be the discrete metric on X. Now suppose that $X' \neq \emptyset$. Define $\rho : X \times X \longrightarrow \mathbb{R}$ as

$$\rho(x, y) = \begin{cases} 0 & \text{if } x = y \\ d(x, y) + \max\{d(x, X'), d(y, X')\} & \text{if } x \neq y. \end{cases}$$

Using the inequality,

$$\max\{u, w\} \leq u + w \leq \max\{u, v\} + \max\{v, w\} \ \forall \ u, \ v, \ w \geq 0,$$

one can easily verify that ρ is a metric on X.

Clearly, every convergent sequence in (X, ρ) is convergent in (X, d). Now let (x_n) be a sequence of distinct points converging to x in (X, d). Then, $x \in X'$ and $d(x, X') = 0$. Thus,

$$\rho(x_n, x) = d(x_n, x) + d(x_n, X')$$

$$\leq 2d(x_n, x).$$

So (x_n) is convergent to x in (X, ρ). Therefore, ρ and d are equivalent metrics on X by Theorem 2.6.1.

Now we use Theorem 5.2.2(a) to prove that (X, ρ) is UC. Let (x_n) and (y_n) be asymptotic sequences in (X, ρ) with $x_n \neq y_n \ \forall \ n \in \mathbb{N}$. Then $0 \leq d(x_n, X') \leq \rho(x_n, y_n) \to 0$ as $n \to \infty$. So there exists a sequence (x'_n) in X' such that $d(x_n, x'_n) \to 0$ as $n \to \infty$. Since X' is compact, by Theorem 4.2.4, (x'_n) has a cluster point in (X, d). Consequently, (x_n) (and equivalently (y_n)) has a cluster point in (X, d). Since d and ρ are equivalent, (x_n) (equivalently (y_n)) has a cluster point in (X, ρ). By Theorem 5.2.2, (X, ρ) is a UC space. $\square$

Remark. The main outline of the proof of Theorem 5.2.3 has been taken from [5].

Recall that the class of Cauchy-continuous functions lies strictly in between the class of uniformly continuous functions and that of continuous functions. Now one might be curious to know when every Cauchy-continuous function on a given metric space is uniformly continuous? Definitely, it should be weaker than UC spaces. In fact, from Corollary 3.5.2 and Exercise 3.5.3, the next result is quite intuitive.

Theorem 5.2.4. *Let* (X, d) *be a metric space and* $(\widehat{X}, d)$ *be its completion. Then the following statements are equivalent:*

(a) *For every metric space* (Y, ρ), *every Cauchy-continuous function* $f : (X, d) \longrightarrow (Y, \rho)$ *is uniformly continuous.*

(b) *Every real-valued Cauchy-continuous function on* (X, d) *is uniformly continuous.*

(c) *The metric space* $(\widehat{X}, d)$ *is a UC space.*

(d) *Every pseudo-Cauchy sequence with distinct terms in* (X, d) *has a Cauchy subsequence.*

We stated the previous result without proof as our main goal behind this result was to make our readers get acquainted with some interesting facts.

Historical Perspective

Probably Raouf Doss [21] was the first one to study UC spaces in 1947 and then it was studied by Jun-iti Nagata [41] in 1950. But UC spaces were first extensively studied by Masahiko Atsuji [2] in 1958. Consequently, UC spaces are also known as *Atsuji spaces*. In recent years, UC spaces have been studied with a different approach by various mathematicians.

Exercises

5.2.1. Let $X = \{(1/j)e_n : n, \ j \in \mathbb{N}\} \cup \{0\} \subseteq l^2$, where $e_n \in l^2$ with 1 at the nth place and 0 otherwise. Show that (X, d) is a UC space where d is the metric induced by the standard metric on l^2. Further, verify that (X, d) is not locally compact.

5.2.2. Think of more examples of UC spaces and non-UC spaces. Can you construct an example of a UC space which is non-compact using Theorem 5.2.3?

5.2.3. Prove that if (x_n) is a sequence in a UC space (X, d) with no cluster point, then the set $A = \{n \in \mathbb{N} : x_n \in X'\}$ is finite.

5.2.4. Give example of a pseudo-Cauchy sequence which does not have a Cauchy subsequence.

5.2.5. Prove that for $x \in X$, $I(x) = \infty$ if and only if $X = \{x\}$. In case, X has at least two points, then show that the isolation functional $I : (X, d) \to [0, \infty)$ is Lipschitz.

5.2.6. Prove Theorem 5.2.4.

5.3   Cofinally Complete Spaces: A Brief Introduction

Recall that a sequence (x_n) is Cauchy if for every $\epsilon > 0$, there exists a residual set of indices $\mathbb{N}_\epsilon$ such that each pair of terms whose indices come from N_ϵ are within ϵ distance from each other. If we replace "residual" by "cofinal" then we obtain sequences that are called cofinally Cauchy. The precise definition is as follows.

Definitions 5.3.1. A sequence (x_n) in a metric space (X, d) is called **cofinally Cauchy** if for each $\epsilon > 0$, there exists an infinite subset N_ϵ of $\mathbb{N}$ such that for each n, $j \in N_\epsilon$, we have $d(x_n, x_j) < \epsilon$.

A metric space (X, d) is said to be **cofinally complete** if every cofinally Cauchy sequence in X clusters.

Example 5.3.1. The set of real numbers $\mathbb{R}$ equipped with the usual metric is cofinally complete: let (x_n) be a cofinally Cauchy sequence in $\mathbb{R}$. Then there exists an infinite subset N of $\mathbb{N}$ such that $|x_n - x_j| < 1$ for all n, $j \in N$. Thus $(x_n)_{n \in N}$ is a bounded sequence in $\mathbb{R}$ and hence by Bolzano–Weierstrass theorem it has a convergent subsequence. This implies that the sequence $(x_n)_{n \in \mathbb{N}}$ clusters in $\mathbb{R}$.

Example 5.3.2. Let $X = \bigcup_{n \in \mathbb{N}} A_n \subseteq l^2$, where $A_n = \{e_n\} \cup \{e_n + \frac{1}{n} e_k : k \in \mathbb{N}\}$ and e_n is the sequence whose nth term is 1 and rest all terms are 0. Then (X, d) is discrete and complete, where 'd' is the metric induced by the metric on l^2 (refer to Example 1.3.7), because every Cauchy sequence in (X, d) is eventually constant. On the other hand, by enumerating X, we will get a cofinally Cauchy sequence which certainly does not cluster and hence (X, d) is not cofinally complete.

Since every Cauchy sequence is cofinally Cauchy, every cofinally complete metric space is complete. In fact, the class of cofinally complete metric spaces lies in between the class of complete metric spaces and that of compact metric spaces. Before moving to the equivalent characterizations of this special class of metric spaces, we would like to mention the following result. It says that every cofinally Cauchy sequence with no constant subsequence has a cofinally Cauchy subsequence of distinct terms.

Proposition 5.3.1. *([7]) Let (x_n) be a cofinally Cauchy sequence in a metric space (X, d) with no constant subsequence. Then there is a pairwise disjoint family $\{M_j : j \in \mathbb{N}\}$ of infinite subsets of $\mathbb{N}$ such that*

(a) *if $\{i, l\} \subseteq \bigcup\{M_j : j \in \mathbb{N}\}$ then $x_i \neq x_l$; and*
(b) *if $i \in M_j$ and $l \in M_j$ then $d(x_i, x_l) < \frac{1}{j}$.*

Proof. Suppose (x_n) has a Cauchy subsequence. Since (x_n) has no constant subsequence, we can assume that (x_n) has a Cauchy subsequence $(x_{n_k})_{k \in \mathbb{N}}$ of distinct terms. Thus, for each $j \in \mathbb{N}$, there exists $m_j \in \mathbb{N}$ such that $d(x_{n_k}, x_{n_l}) < \frac{1}{j}$ for all k, $l \geq m_j$. Let $\mathbb{N}_o = \{n_k : k \in \mathbb{N}\}$. Then partition $\mathbb{N}_o$ into countably many infinite subsets $\{\mathbb{K}_j : j \in \mathbb{N}\}$. Now for each $j \in \mathbb{N}$, choose $\mathbb{M}_j = \{n_k \in \mathbb{K}_j : k \geq m_j\}$.

If (x_n) has no Cauchy subsequence, choose an infinite subset $\mathbb{M}_1$ of $\mathbb{N}$ such that $0 < d(x_i, x_l) < 1$ for all i, $l \in \mathbb{M}_1$. Since (x_n) has no Cauchy subsequence, the set $\{x_i : i \in \mathbb{M}_1\}$ cannot be totally bounded. By passing to an infinite subset of $\mathbb{M}_1$, we can find $\epsilon_1 < \frac{1}{2}$ such that $\epsilon_1 < d(x_i, x_l) < 1$ for all i, $l \in \mathbb{M}_1$. Now choose an infinite subset $\mathbb{M}_2$ of $\mathbb{N}$ such that $0 < d(x_i, x_l) < \epsilon_1$ for all i, $l \in \mathbb{M}_2$. By construction $\{x_i : i \in \mathbb{M}_1\} \cap \{x_i : i \in \mathbb{M}_2\}$ consists of at most one point. Also, $\{x_i : i \in \mathbb{M}_2\}$ is not totally bounded, so by passing to an infinite subset of $\{x_i : i \in \mathbb{M}_2\}$ we can assume the two sets are disjoint and further that there exists $\epsilon_2 < \frac{1}{3}$ such that $\epsilon_2 < d(x_i, x_l) < \frac{1}{2}$ for all i, $l \in \mathbb{M}_2$. Choosing an infinite $\mathbb{M}_3 \subseteq \mathbb{N}$ such that $0 < d(x_i, x_l) < \epsilon_2$ for all i, $l \in \mathbb{M}_3$. By deleting at most two indices from $\mathbb{M}_3$ we can assume $\{\{x_i : i \in \mathbb{M}_j\} : j = 1, 2, 3\}$ is a pairwise disjoint family. Continuing in this way inductively we produce $\{\mathbb{M}_j : j \in \mathbb{N}\}$ with the required properties. $\square$

As a consequence of the previous result, it is enough to show the clustering of cofinally Cauchy sequences of *distinct* terms in order to prove a space to be cofinally complete. And hence by the sequential characterization of UC spaces in terms of pseudo-Cauchy sequences (Theorem 5.2.2), it is evident that the class of cofinally complete metric spaces is positioned strictly between the class of UC spaces and that of complete metric spaces. Note that the set of real numbers, $\mathbb{R}$, with the usual metric is cofinally complete but not UC.

Recall that a metric space (X, d) is called a UC space if every real-valued continuous function on (X, d) is uniformly continuous. Since every UC space is cofinally complete, one may think of having a similar characterization of cofinally complete metric spaces. For that let us define the following class of functions.

Definition 5.3.2. A function $g : (X, d) \to (Y, \rho)$ between two metric spaces is called **uniformly locally bounded** if $\exists$ a $\delta > 0$ such that $\forall\, x \in X$, $g(B_d(x, \delta))$ is a bounded subset of (Y, ρ).

Evidently, every uniformly continuous function is uniformly locally bounded. But it should be noted that a continuous function may not be uniformly locally bounded in general. For example, let $f : (0,1) \to \mathbb{R}$ defined by $f(x) = 1/x$. The next result gives a necessary and sufficient condition on a metric space (X,d) such that every continuous function defined on (X,d) is uniformly locally bounded.

Theorem 5.3.1 ([7]). *Let (X,d) be a metric space. Then the following statements are equivalent:*

(a) *(X,d) is cofinally complete.*
(b) *Each continuous function on (X,d) with values in an arbitrary metric space (Y,ρ) is uniformly locally bounded.*
(c) *Each real-valued continuous function on (X,d) is uniformly locally bounded.*

Proof. (a) $\Rightarrow$ (b): Let $f : (X,d) \to (Y,\rho)$ be a continuous function which fails to be uniformly locally bounded. Then there exists a sequence (x_n) in X such that $f(B_d(x_n, \frac{1}{n}))$ is an unbounded subset of (Y,ρ). Thus by Theorem 4.3.1 and Proposition 4.1.2, the closed ball $C_d(x_n, \frac{1}{n})$ is not compact. We claim that the sequence (x_n) does not cluster in (X,d). Suppose, if possible, $x \in X$ is a cluster point of (x_n). Since f is continuous at x, there exists a $\delta > 0$ such that $f(B_d(x,\delta)) \subseteq B_\rho(f(x),1)$. By Theorem 1.6.1, there exists a subsequence (x_{n_k}) of (x_n) which converges to x. Then for sufficiently large k, $B_d(x_{n_k}, \frac{1}{n_k}) \subseteq B_d(x,\delta)$ and hence $f(B_d(x_{n_k}, \frac{1}{n_k})) \subseteq f(B_d(x,\delta)) \subseteq B_\rho(f(x),1)$. This gives a contradiction. Thus, (x_n) has no cluster point in X.

Since $C_d(x_n, \frac{1}{n})$ is not compact, let $(a_m^n)_{m \in \mathbb{N}}$ be a sequence of distinct points in $C_d(x_n, \frac{1}{n})$ with no cluster point (Theorem 4.2.4). Then $B = \{x_n, a_m^n : m,\ n \in \mathbb{N}\}$ is a closed and discrete subset of X. Since B is countable, let $\{z_n : n \in \mathbb{N}\}$ be an enumeration of B. Consider the function $f : B \to \mathbb{R}$: $f(z_n) = n \quad - (\star)$. Then f is a continuous function on a closed set B and hence by Tietze's extension theorem, f can be extended to a real-valued continuous function $\widetilde{f}$ on (X,d). Now (z_n) is cofinally Cauchy in X and (X,d) is cofinally complete, hence (z_n) has a convergent subsequence. Then by the continuity of $\widetilde{f}$, $(\widetilde{f}(z_n))$ has a convergent subsequence. This gives a contradiction to $(\star)$. Consequently, f is uniformly locally bounded.

(b) $\Rightarrow$ (c): This is immediate.

(c) $\Rightarrow$ (a): If (X,d) is not cofinally complete, then there exists a cofinally Cauchy sequence (x_n) of distinct points in (X,d) with no cluster point. Thus the set $A = \{x_n : n \in \mathbb{N}\}$ is closed and discrete and hence the function $f : (A,d) \to \mathbb{R}$ defined by $f(x_n) = n$ is continuous. By Tietze's extension theorem, there exists a real-valued continuous extension $\widetilde{f}$ of f to (X,d). By (c), $\widetilde{f}$ is uniformly locally bounded and hence $\exists$ a $\delta > 0$ such that $\forall\ x \in X$, $\widetilde{f}(B(x,\delta))$ is a bounded subset of $\mathbb{R}$ ($\star\star$). Since (x_n) is cofinally Cauchy, there exists an infinite subset N of $\mathbb{N}$ and $n_o \in \mathbb{N}$ such that $x_m \in B(x_{n_o},\delta)$ for all $m \in N$. Then the set $S = \{\widetilde{f}(x_m) : m \in N\} \subseteq \widetilde{f}(B(x_{n_o},\delta))$ is bounded by ($\star\star$). This is a contradiction to the way f is defined. Thus (X,d) is cofinally complete. $\square$

Since a cofinally complete metric space is complete, we now look for some nice characterizations of the metric spaces whose completions are cofinally complete.

Theorem 5.3.2. *Let (X,d) be a metric space. Then the following statements are equivalent:*

(a) *The completion $(\widehat{X},d)$ of (X,d) is cofinally complete.*
(b) *Every complete subset (as a metric subspace) of (X,d) is cofinally complete.*
(c) *Every cofinally Cauchy sequence in (X,d) has a Cauchy subsequence.*
(d) *Each Cauchy-continuous function on (X,d) with values in an arbitrary metric space (Y,ρ) is uniformly locally bounded.*
(e) *Each real-valued Cauchy-continuous function on (X,d) is uniformly locally bounded.*

Proof. The implications (a) $\Rightarrow$ (b); (d) $\Rightarrow$ (e) are easy to see.

(b) $\Rightarrow$ (c): Let (x_n) be a cofinally Cauchy sequence in X. We claim that (x_n) has a Cauchy subsequence. Suppose (x_n) has no Cauchy subsequence. Then $A = \{x_n : n \in \mathbb{N}\}$ is complete as a metric subspace of (X,d). Hence (A,d) is cofinally complete, which implies that (x_n) has a cluster point. This is a contradiction as (x_n) has no Cauchy subsequence.

(c) $\Rightarrow$ (a): Let $(\widehat{x}_n)$ be a cofinally Cauchy sequence in $(\widehat{X}, d)$. Then by the density of X in $\widehat{X}$, for every $n \in \mathbb{N}$, there exists $x_n \in X$ such that $d(x_n, \widehat{x}_n) < 1/n$. Then (x_n) is cofinally Cauchy in (X, d) and hence it has a Cauchy subsequence, say (x_{n_k}). Thus, (x_{n_k}) converges in $(\widehat{X}, d)$. Consequently, we have a convergent subsequence of $(\widehat{x}_n)$.

(a) $\Rightarrow$ (d): Let $f : (X, d) \to (Y, \rho)$ be a Cauchy-continuous function. Then by Theorem 3.5.3, there exists a Cauchy-continuous function $\widetilde{f} : (\widehat{X}, d) \to (\widehat{Y}, \rho)$ which extends f. Now by Theorem 5.3.1, $\widetilde{f}$ is uniformly locally bounded. Hence, f is uniformly locally bounded.

(e) $\Rightarrow$ (a): Let $f : (\widehat{X}, d) \to \mathbb{R}$ be a continuous function. We claim that f is uniformly locally bounded. Since $(\widehat{X}, d)$ is complete, f is Cauchy-continuous (Corollary 3.5.2) which implies that the restriction of f to X, $f|_X$, is Cauchy continuous and hence uniformly locally bounded (by (e)). Then there exists a $\delta > 0$ such that $\forall\, x \in X$, $f(B(x, \delta) \cap X)$ is bounded. Let $\widehat{x} \in \widehat{X}$. We claim that $f(B(\widehat{x}, \frac{\delta}{3}))$ is bounded. Suppose it is not bounded. Then there exists a sequence $(\widehat{x}_n)$ in $B(\widehat{x}, \frac{\delta}{3})$ such that $|f(\widehat{x}_n)| > n$. Let $(a_m^n)_{m \in \mathbb{N}}$ be a sequence in X converging to $\widehat{x}_n$ for every $n \in \mathbb{N}$. Then, eventually every such sequence lies in some δ ball of X and hence there images under f are bounded by some $M > 0$. Subsequently, by the continuity of f, $(f(\widehat{x}_n))$ is bounded, which is a contradiction to the way $(\widehat{x}_n)$ is constructed. $\qquad\square$

Historical Perspective

Cofinal completeness was first considered implicitly by Corson [19] in 1958 and then by Howes [29] in 1971 in terms of nets and entourages. A few years later, Rice [45] introduced the notion of uniform para-compactness for a Hausdorff uniform space X and subsequently in Ref. [53], Smith, the reviewer of Rice's paper for *Mathematical Reviews*, observed that uniform paracompactness is equivalent to net cofinal completeness for a Hausdorff uniform space. In 1981, Hohti [28] gave a nice equivalent characterization of a uniformly paracompact metric space in terms of uniform local compactness. Much later in 2008, Beer [7] cast a new light on cofinal complete-ness and gave various nice characterizations of cofinally complete metric spaces. Besides this, Beer also discussed some equivalent con-ditions under which a metric space possesses an equivalent cofinally complete metric. For a detailed discussion on cofinally complete metric spaces, the interested readers can also have a look at the research monograph [32].

Exercises

5.3.1. Give an example of

(a) a complete metric space which is not cofinally complete;
(b) a cofinally Cauchy sequence with no Cauchy subsequence.

5.3.2 ([7]). Prove that the following assertions are equivalent for a metric space (X, d):

(a) (X, d) is totally bounded.
(b) Each sequence in X is cofinally Cauchy.
(c) Each sequence in X is pseudo-Cauchy.

5.3.3. Let (X, d) be a metric space. Then prove that the following statements are equivalent:

(a) (X, d) is cofinally complete.
(b) Every d-cofinally Cauchy sequence (that is, cofinally Cauchy with respect to metric d) in X is σ-cofinally Cauchy for all equivalent metrics σ on X.

5.3.4. Let us define analogues of asymptotic sequences and uniformly asymptotic sequences: a pair of sequences (x_n) and (y_n) in a metric space (X, d) is said to be

(i) *cofinally asymptotic*, written $(x_n) \asymp^c (y_n)$, if $\forall\, \epsilon > 0$, $\exists$ an infinite subset N_ϵ of $\mathbb{N}$ such that $d(x_n, y_n) < \epsilon \,\forall\, n \in N_\epsilon$.
(ii) *cofinally uniformly asymptotic*, written $(x_n) \asymp^c_u (y_n)$, if $\forall\, \epsilon > 0$, $\exists$ an infinite subset N_ϵ of $\mathbb{N}$ such that $d(x_n, y_m) < \epsilon \,\forall\, n,\, m \in N_\epsilon$.

Now suppose that (x_n) and (y_n) are sequences in a metric space (X, d) and $b \in X$. Show that

(a) If $(x_n) \asymp^c (y_n)$, then there exist subsequences $(x_{n_k})_{k \in \mathbb{N}}$ and $(y_{n_k})_{k \in \mathbb{N}}$ of (x_n) and (y_n) respectively, such that $(x_{n_k}) \asymp (y_{n_k})$.
(b) A sequence (x_n) has a subsequence converging to b if and only if $(x_n) \asymp^c_u (b)_{n \in \mathbb{N}}$ if and only if $(x_n) \asymp^c (b)_{n \in \mathbb{N}}$.
(c) For every sequence (x_n) in X, $(x_n) \asymp^c (x_n)$.

(d) A sequence (x_n) is cofinally Cauchy if and only if $(x_n) \asymp_u^c (x_n)$.
(e) If $(x_n) \asymp_u^c (y_n)$, then (x_n) and (y_n) are cofinally Cauchy sequences.
(f) If (x_n) is cofinally Cauchy such that $(x_n) \asymp (y_n)$ then $(x_n) \asymp_u^c (y_n)$.
(g) If (x_n) is a Cauchy sequence such that $(x_n) \asymp^c (y_n)$ then $(x_n) \asymp_u^c (y_n)$.

5.3.5. Let $f : (X, d) \to (Y, \rho)$ be a function between two metric spaces. Then prove that the following statements are equivalent:

(a) f is uniformly continuous.
(b) Whenever $(x_n) \asymp^c (z_n)$ in (X, d), then $(f(x_n)) \asymp^c (f(z_n))$ in (Y, ρ).
(c) Whenever $(x_n) \asymp (z_n)$ in (X, d), then $(f(x_n)) \asymp^c (f(z_n))$ in (Y, ρ).

5.4 Finite Chainability

Recall that a metric space (X, d) is compact if and only if every real-valued continuous function on (X, d) is bounded. Now the analogous question arises for boundedness of real-valued uniformly continuous functions. In this regard, first note that every real-valued uniformly continuous function on a totally bounded metric space (X, d) is bounded (Corollary 3.2.4). But the converse need not be true.

Example 5.4.1. Consider the space of all real square summable sequences l^2 (see Example 1.3.7). Then the open ball around the zero sequence $\mathbf{0}$ in l^2, $B(\mathbf{0}, 2)$, is not totally bounded: consider the sequence $(e_n) \subseteq B(\mathbf{0}, 2)$, where e_n is the sequence with 1 as nth term and rest are 0. Then $d(e_n, e_k) = \sqrt{2}$ for $n \neq k$ and hence (e_n) has no Cauchy subsequence. Now see Theorem 3.2.3. But note that every real-valued uniformly continuous function on $B(\mathbf{0}, 2)$ is bounded: let $f : B(\mathbf{0}, 2) \to \mathbb{R}$ be uniformly continuous. Then there exists a $\delta > 0$ such that $|f(x) - f(y)| < 1$ whenever $x, y \in B(\mathbf{0}, 2)$ with $d(x, y) < \delta$. Let $m \in \mathbb{N}$ such that $\frac{1}{m} < \frac{\delta}{2}$. If $x \in B(\mathbf{0}, 2)$, then we can find m points in $B(\mathbf{0}, 2)$, $x_0 = \mathbf{0}$, $x_1, \ldots, x_{m-2}, x_{m-1} = x$, such that $d(x_{i-1}, x_i) < \delta$ for all $i \in \{1, \ldots, m-1\}$ (take $x_i = \frac{i}{m} x$).

Consequently, $|f(x)-f(\mathbf{0})| \leq |f(x)-f(x_{m-2})|+\cdots+|f(x_1)-f(\mathbf{0})| < m-1$ for all $x \in B(\mathbf{0},2)$. Thus, f is bounded.

In order to study such metric spaces on which every real-valued uniformly continuous function is bounded, in 1958 Atsuji [2] introduced finitely chainable metric spaces which were weaker than totally bounded metric spaces but stronger than bounded metric spaces. In fact, one can get the intuition for the definition by Example 5.4.1.

Definitions 5.4.1. Let (X,d) be a metric space and ϵ be a positive number. Then an ordered set of points $\{x_0, x_1, \ldots, x_m\}$ in X satisfying $d(x_{i-1}, x_i) \leq \epsilon$, where $i = 1, 2, \ldots, m$, is said to be an ϵ-*chain of length m* from x_o to x_m.

Let A be a subset of X. Then, A is said to be **finitely chainable** in (X,d) if for every $\epsilon > 0$, there exist finitely many points $p_1, p_2, \ldots, p_r$ in X and a positive integer m such that every point of A can be joined with some p_j, $1 \leq j \leq r$ by an ϵ-chain of length m.

Remark. These finitely chainable sets are also known as *Bourbaki-bounded* sets in the literature because these sets were considered in the book of Bourbaki [12].

It is evident that every subset of a finitely chainable set in a metric space (X,d) is also finitely chainable in X. Here we would like to point out that unlike total boundedness, finite chainability of a set A depends essentially on the underlying space (X,d), that is, if Y is a subset of (X,d), then $A \subseteq Y$ is finitely chainable in X whenever it is finitely chainable in Y, but the converse need not hold. For example, the set $A = \{e_n : n \in \mathbb{N}\}$ in l^2 is finitely chainable in the whole space l^2: for every $\epsilon > 0$, we can find $m \in \mathbb{N}$ such that every e_n can be joined with the zero sequence by some ϵ-chain of length m. But A is not finitely chainable subset of itself as A is an infinite uniformly discrete metric space.

The routine proof of the following result is omitted.

Proposition 5.4.1. *Let (X,d) be a metric space. Then*

(a) *Every totally bounded subset of (X,d) is finitely chainable in (X,d).*
(b) *Every finitely chainable subset of (X,d) is bounded.*
(c) *If A is finitely chainable in (X,d), then $\overline{A}$ is also finitely chainable in (X,d).*

Observe that a bounded subset of a metric space (X, d) need not be finitely chainable in (X, d). For example, consider any infinite subset of a metric space (X, d), where d is the discrete metric. But interestingly, this cannot be the case in the Euclidean space $\mathbb{R}^n$ and in l^2 (see Exercise 5.4.1). Furthermore, it should be noted that the open ball $B(\mathbf{0}, 2)$ and the set $A = \{e_n : n \in \mathbb{N}\}$ (see Example 5.4.1) are finitely chainable in l^2 but both are not totally bounded.

Now we prove the main result of this section due to which the notion of finite chainability was introduced.

Theorem 5.4.1. *Let A be a subset of the metric space (X, d). Then the following statements are equivalent:*

(a) *A is finitely chainable in (X, d).*
(b) *Every uniformly continuous function defined on (X, d) with values in an arbitrary metric space (Y, ρ) is bounded on A.*
(c) *Every real-valued uniformly continuous function on (X, d) is bounded on A.*

Proof. (a) $\Rightarrow$ (b): Let $f : (X, d) \to (Y, \rho)$ be uniformly continuous. Then there exists a $\delta > 0$ such that $\rho(f(x), f(z)) < 1$ whenever $d(x, z) < \delta$. Since A is finitely chainable in (X, d), there exist finitely many points $p_1, p_2, \ldots, p_r$ in X and $m \in \mathbb{N}$ such that every point of A can be joined with some p_j, $1 \le j \le r$ by some δ-chain of length m. Thus, for every $a \in A$, $\rho(f(a), f(p_i)) < m$ for some $i \in \{1, 2, \ldots, r\}$. Hence f is bounded on A.

(b) $\Rightarrow$ (c): Trivial.

(c) $\Rightarrow$ (a): Suppose A is not finitely chainable in (X, d). Then there exists an $\epsilon_o > 0$ such that for any finitely many points in X and $n \in \mathbb{N}$, there is a point in A which cannot be joined with any of the points in the finite collection by an ϵ_o-chain of length n. Now fix $x_o \in X$. Let $A_o^o = \{x_o\}$ and for $n \in \mathbb{N}$, let

$A_o^n = \{x \in X : x$ can be joined with x_o by ϵ_o-chain of length n

but not $n - 1\}$.

Now define $f : (X, d) \to \mathbb{R}$ as follows:

Case I: $A \cap A_o^n$ is non-empty for infinitely many n.

If $x \in A_o^n$ for some $n \in \mathbb{N}$, then $f(x) = (n-1)\epsilon_o + d(x, A_o^{n-1})$, otherwise $f(x) = 0$. We claim that f is uniformly continuous. Let $\epsilon < \epsilon_o$ and suppose $d(x, y) < \epsilon$. If $x \notin A_o^n$ for all $n \in \mathbb{N} \cup \{0\}$, then so does y and hence $f(x) = f(y) = 0$. Now suppose $x \in A_o^n$ for some $n \geq 2$. Then we can have the following cases:

Case (a): $y \in A_o^{n-1}$.

Then $d(x, A_o^{n-1}) \leq d(x, y) < \epsilon$ and $d(y, A_o^{n-2}) < \epsilon_o$. This implies that $f(x) - f(y) = \epsilon_o + d(x, A_o^{n-1}) - d(y, A_o^{n-2}) > 0$. Now we claim that $d(y, A_o^{n-2}) \geq \epsilon_o - \epsilon$. Suppose, if possible, $d(y, A_o^{n-2}) < \epsilon_o - \epsilon$. Then there exists $y' \in A_o^{n-2}$ such that $d(y, y') < \epsilon_o - \epsilon$. Thus, $d(x, y') \leq d(x, y) + d(y, y') < \epsilon_o$, which means either $x \in A_o^{n-1}$ or $x \in A_o^{n-3}$. A contradiction. Consequently, $f(x) - f(y) < \epsilon_o + \epsilon + \epsilon - \epsilon_o = 2\epsilon$.

Case (b): $y \in A_o^n$.

Then $|f(x) - f(y)| \leq d(x, y) < \epsilon$.

Case (c): $y \in A_o^{n+1}$.

It can be handled in a way similar to Case (a). Hence, f is a uniformly continuous function which is unbounded on A. This is a contradiction.

Case II: $A \cap A_o^n$ is empty eventually.

Since A is not finitely chainable in (X, d), we can find $a_1 \in A$ such that $a_1 \notin A_o = \bigcup_{n=0}^{\infty} A_o^n$. If we can define a function similar to that in Case I corresponding to a_1, then we are done. In case we are not able to define a function similar to that in Case I for some a_i, since A is not finitely chainable in (X, d), we have a sequence of infinitely many sets $A_o, A_1, A_2, \ldots$. Now define the function f on X as follows: $f(x) = n$ if $x \in A_n$ and $f(x) = 0$ if $x \notin \bigcup_{n=0}^{\infty} A_n$. Then f is a uniformly continuous function which is unbounded on A. Again a contradiction! $\square$

Remark. Note that Theorem 5.4.1 was initially proved by Atsuji [2], particularly for $A = X$. Then Hejcman gave a more general result for finite chainability of subsets of a uniform space in Ref. [26]. The result was later proved in metric space setting by Marino, Lewicki and Pietramala in Ref. [37] and separately by Beer and Garrido in Ref. [9].

The next result is straightforward from Theorem 4.3.4 and the previous result.

Corollary 5.4.1. *A metric space (X, d) is compact if and only if (X, d) is finitely chainable and a UC space.*

It is well known from calculus that the pointwise product of two real-valued continuous functions is also continuous. But this is not in general true for uniformly continuous functions: suppose $f : \mathbb{R} \to \mathbb{R}$ is the identity function, that is, $f(x) = x \; \forall \, x \in \mathbb{R}$. Then f is uniformly continuous but f^2 is not (see Example 2.3.2). Here note that f is unbounded.

Corollary 5.4.2. *Let f and g be two real-valued uniformly continuous functions defined on a finitely chainable metric space (X, d). Then their pointwise product $f.g$ is also uniformly continuous.*

Proof. Since (X, d) is finitely chainable, the uniformly continuous functions f and g are bounded by Theorem 5.4.1. So let M be a positive number such that $|f(x)| \leq M$ and $|g(x)| \leq M$ for all $x \in X$. Let $\epsilon > 0$. Then there exists a $\delta_1 > 0$ such that whenever $d(x, y) < \delta_1$ then $|f(x) - f(y)| < \frac{\epsilon}{2M}$. Similarly, there exists a $\delta_2 > 0$ such that whenever $d(x, y) < \delta_2$ then $|g(x) - g(y)| < \frac{\epsilon}{2M}$. Now suppose $\delta = \min\{\delta_1, \delta_2\}$. Then $\delta > 0$. If $d(x, y) < \delta$, then

$$
\begin{aligned}
|f(x)g(x) - f(y)g(y)| &= |f(x)g(x) - f(x)g(y) + f(x)g(y) - f(y)g(y)| \\
&\leq |f(x)||g(x) - g(y)| + |g(y)||f(x) - f(y)| \\
&< M\frac{\epsilon}{2M} + M\frac{\epsilon}{2M} = \epsilon.
\end{aligned}
$$

Thus, $f \cdot g$ is uniformly continuous. $\qquad\qquad\square$

Note that the converse of the previous result may not hold true, that is, if the class of real-valued uniformly continuous functions on (X, d) is closed under pointwise product, then (X, d) may not be finitely chainable. For example, consider any infinite uniformly discrete metric space. This example gives an idea for the next result.

Theorem 5.4.2 (Cabello Sánchez theorem [15]). *Let (X, d) be a metric space. Then the following statements are equivalent:*

(a) *If f and g are real-valued uniformly continuous functions on (X, d), then their pointwise product $f \cdot g$ is also uniformly continuous.*

(b) *Every subset A of X is either finitely chainable in (X, d) or A contains an infinite uniformly isolated subset (that is, $\exists\ \delta > 0$ and an infinite subset $F \subseteq A$ such that $d(w, X \setminus \{w\}) \geq \delta\ \forall\ w \in F$).*

Proof. (a) $\Rightarrow$ (b): Suppose, if possible, there exists $A \subseteq X$ which is neither finitely chainable nor contains an infinite uniformly isolated subset. Then there exists a uniformly continuous function $f : (X, d) \to \mathbb{R}$ which is unbounded on A (see Theorem 5.4.1). Without loss of generality, we can assume that $f \geq 0$. Consequently, there exists a sequence $(x_n) \subseteq A$ such that $f(x_{n+1}) \geq f(x_n) + 1 \geq n + 1$ for all $n \in \mathbb{N}$. Thus, $|f(x_n) - f(x_m)| \geq 1\ \forall\ n \neq m$. Since f is uniformly continuous, there exists a $\delta > 0$ such that $d(x_n, x_m) \geq \delta\ \forall\ n \neq m$ $(\star)$. Since A does not contain any infinite uniformly isolated subset, for all $\delta' > 0$, the set $\{a \in A : d(a, X \setminus \{a\}) \geq \delta'\}$ is finite. Consequently, $d(x_n, X \setminus \{x_n\}) \to 0$ and hence we can find a sequence (y_n) in X such that for all $n \in \mathbb{N}$, $0 < d(x_n, y_n) < \delta/3$ and $d(x_n, y_n) \to 0$. Note that by $(\star)$, the sequence $\langle x_1, y_1, x_2, y_2, \ldots \rangle$ has distinct terms. Now define a function $g : \{x_n, y_n : n \in \mathbb{N}\} \to [0, 1]$ as: $g(x_n) = 1/n$ and $g(y_n) = 0$. Then g is uniformly continuous. By Theorem 3.5.5, g can be extended to a uniformly continuous function $\widetilde{g} : (X, d) \to [0, 1]$. Since for all n $f \cdot \widetilde{g}(x_n) \geq 1$ and $f \cdot \widetilde{g}(y_n) = 0$, $f \cdot \widetilde{g}$ is not uniformly continuous. A contradiction!

(b) $\Rightarrow$ (a): Let $f,\ g : (X, d) \to \mathbb{R}$ be uniformly continuous functions whose pointwise product $f \cdot g$ is not uniformly continuous. Then there exists an $\epsilon_o > 0$ such that for each $n \in \mathbb{N}$, we can find $x_n,\ y_n \in X$ with $0 < d(x_n, y_n) < 1/n$ but $|f(x_n)g(x_n) - f(y_n)g(y_n)| \geq \epsilon_o$. Consequently, either f or g is unbounded on the set $A = \{x_n, y_n : n \in \mathbb{N}\}$. Then A is not finitely chainable in (X, d) by Theorem 5.4.1. Since $0 < d(x_n, y_n) < 1/n$, $d(x_n, X \setminus \{x_n\}) < 1/n$ and $d(y_n, X \setminus \{y_n\}) < 1/n$. Thus, for all $\delta > 0$, the set $\{a \in A : d(a, X \setminus \{a\}) \geq \delta\}$ is finite. This implies that A does not contain any infinite uniformly isolated subset, which is again a contradiction. $\square$

Like we have Cauchy sequences which characterize totally bounded metric spaces, similarly in 2014, Garrido and Meroño [23] defined a special class of sequences called Bourbaki–Cauchy to give nice sequential characterizations of finitely chainable metric spaces. We give the precise definition first.

Definition 5.4.2. Let (X, d) be a metric space. A sequence (x_n) is said to be **Bourbaki–Cauchy** in X if for every $\epsilon > 0$, there exist m, $n_o \in \mathbb{N}$ such that whenever $n > j \geq n_o$, the points x_j and x_n can be joined by an ϵ-chain of length m.

Clearly, every Cauchy sequence is Bourbaki–Cauchy but the reverse implication need not hold true. For example, the sequence (e_n) is Bourbaki–Cauchy in l^2 but it is not Cauchy. Here note that (e_n) is not Bourbaki–Cauchy in the subset $X = \{e_n : n \in \mathbb{N}\} \subseteq l^2$. Hence unlike Cauchy sequences, the ambient space matters when we talk about Bourbaki–Cauchy sequences. Interestingly, finitely chainable metric spaces are characterized by Bourbaki–Cauchy sequences in the same way as Cauchy sequences characterize total boundedness (see Theorem 3.2.3).

Theorem 5.4.3. *Let (X, d) be a metric space and $A \subseteq X$. Then the following statements are equivalent:*

(a) *A is finitely chainable in X.*
(b) *Every countable subset of A is finitely chainable in X.*
(c) *Every sequence in A has a Bourbaki–Cauchy subsequence in X.*

Proof. (a) $\Rightarrow$ (b): This is immediate.

(b) $\Rightarrow$ (c): Let (x_n) be a sequence in A. Then $\{x_n : n \in \mathbb{N}\}$ is finitely chainable in X. Thus, there exists $p_1 \in X$ and $m_1 \in \mathbb{N}$ and a subsequence (x_n^1) of (x_n) such that every point of the subsequence can be joined with p_1 by 1-chain of length m_1. Repeating this process we get, for every $k \geq 2$ there exist $p_k \in X$ and $m_k \in \mathbb{N}$ and a subsequence (x_n^k) of (x_n^{k-1}) such that every point of the subsequence can be joined with p_k by $\frac{1}{k}$-chain of length m_k. Then the sequence (x_n^n) is the required Bourbaki–Cauchy subsequence of (x_n).

(c) $\Rightarrow$ (a): Suppose A is not finitely chainable in X. Then there exists an $\epsilon_o > 0$ such that for any finitely many points in X and $m \in \mathbb{N}$, there is a point in A which cannot be joined with any of the points in the finite collection by an ϵ_o-chain of length m. Now fix $x_0 \in X$. Then for all $m \in \mathbb{N}$, there exists $a_m \in A$ such that a_m cannot be joined with $x_0, a_1, \ldots, a_{m-1}$ by any ϵ_o-chain of length m. Consequently, the sequence $(a_m)_{m \in \mathbb{N}}$ in A has no Bourbaki–Cauchy subsequence in X, which is a contradiction. $\qquad\square$

One can find a wide collection of equivalent characterizations of finitely chainable metric spaces in the expository article [33]. Moreover, for further study of UC spaces, cofinal completeness and finite chainability, one may refer to the research monograph [8].

Exercises

5.4.1. Show that in the Euclidean space $\mathbb{R}^n$ and in l^p space, a subset A is bounded if and only if it is finitely chainable in the whole space.

5.4.2. Prove Proposition 5.4.1.

5.4.3. Prove that a metric space (X, d) is finitely chainable if and only if (X, ρ) is bounded for every metric ρ which is uniformly equivalent to the metric d.

5.4.4. Let $X = \bigcup_{n \in \mathbb{N}} A_n$ be the discrete subset of l^2, where $A_n = \{e_n, \ e_n + \frac{1}{n} e_k : k \in \mathbb{N}\}$. What can you say about Bourbaki–Cauchy sequences in X?

Chapter 6

Real-Valued Functions on Metric Spaces

This chapter is devoted to the analysis of real-valued functions defined on a metric space. Particularly, we will study three significant results, namely Dini's Theorem, Ascoli–Arzelà Theorem and Stone–Weierstrass Theorem. These results play a crucial role in the analysis of the metric space $(C(X), D)$ where $C(X)$ denotes the set of all real-valued continuous functions on a compact metric space (X, d) and D is the uniform metric.

6.1 Pointwise and Uniform Convergence

We are already familiar with the significance of the study of convergence of sequences in real analysis. Now the question is: *How to define the concept of convergence of a sequence of real-valued functions?* So let us start with a sequence (f_n) of real-valued functions defined on a non-empty set X. We know that for a fixed $x \in X$, $(f_n(x))$ is a sequence of real numbers. Intuitively, if we want the sequence (f_n) of functions to converge, then the sequence $(f_n(x))$ should converge in $(\mathbb{R}, |\cdot|)$ for all $x \in X$ (that is, $\lim\limits_{n \to \infty} f_n(x)$ exists in $\mathbb{R}$ for each $x \in X$). Consequently, by collecting all the limits with respect to each point in X, we can formulate a new function $f : X \to \mathbb{R}$. Thus, f is defined by $f(x) = \lim\limits_{n \to \infty} f_n(x)$ for $x \in X$. If this happens, then the sequence (f_n) is said to converge pointwise to

f (or f is the pointwise limit of (f_n)). So we can write the formal definition of pointwise convergence as follows:

Definition 6.1.1. Let (f_n) be a sequence of real-valued functions defined on a non-empty set X and let $f : X \to \mathbb{R}$ be a function. Then we say that (f_n) converges to f **pointwise**, denoted by $f_n \to f$ pointwise, if given $\epsilon > 0$ and $x \in X$, there exists $n_o \in \mathbb{N}$ (n_o depends on both ϵ and x) such that

$$|f_n(x) - f(x)| < \epsilon \ \ \forall \, n \geq n_o.$$

In the previous definition, it should be noted that n_o depends on both ϵ and x. Suppose n_o depends only on ϵ and not on x, then we get a stronger notion of convergence of a sequence of real-valued functions. This stronger concept is known as *uniform convergence* because the natural number n_o is uniform for all points in the domain space (X, d). The formal definition is as follows.

Definition 6.1.2. Let (f_n) be a sequence of real-valued functions defined on a non-empty set X. Then we say that (f_n) converges to $f : X \to \mathbb{R}$ **uniformly**, denoted by $f_n \to f$ uniformly, if given $\epsilon > 0$, there exists $n_o \in \mathbb{N}$ (n_o depends on ϵ only) such that

$$|f_n(x) - f(x)| < \epsilon \ \ \forall \, n \geq n_o \text{ and } \forall \, x \in X.$$

Remarks. (a) When a sequence (f_n) of real-valued functions converges uniformly to a function f, we call f to be 'the' uniform limit of (f_n). (Here the article 'the' is used because of the uniqueness of limit of a sequence, see Proposition 1.5.1.)

(b) The concept of pointwise and uniform convergence can also be defined for a sequence of functions with codomain as some general metric space but we focus on real-valued functions in this chapter.

(c) A sequence $(f_n) \subseteq \mathbb{R}^X$ is said to converge uniformly on $K \subseteq X$ to the function f in $\mathbb{R}^X$ if for each $\epsilon > 0$, there exists $n_o \in \mathbb{N}$ such that

$$|f_n(x) - f(x)| < \epsilon \, \forall \, x \in K \quad \text{and} \quad \forall \, n \geq n_o,$$

which is equivalent to $\lim_{n \to \infty} (\sup_{x \in K} |f_n(x) - f(x)|) = 0$.

Evidently, uniform convergence implies pointwise convergence and hence by the uniqueness of limits (Proposition 1.5.1), the uniform limit is also the pointwise limit. Consequently, whenever we are given a sequence of functions then first find out its pointwise limit and subsequently use this pointwise limit to check whether the convergence is uniform or not.

Example 6.1.1. Consider the sequence of functions (f_n) where $f_n : [0, \infty) \to \mathbb{R}$ is defined as $f_n(x) = xe^{-nx}$ for $x \in [0, \infty)$. First see that $\lim_{n \to \infty} f_n(x) = 0$ for each $x \in [0, \infty)$. Thus, the zero function is the pointwise limit of (f_n). Now we show that $f_n \to 0$ uniformly. Let $\epsilon > 0$. We need to find $n_o \in \mathbb{N}$ such that $|f_n(x) - f(x)| < \epsilon \ \forall n \geq n_o$ and $\forall x \in [0, \infty)$, which is equivalent to $\sup_{x \in [0,\infty)} xe^{-nx} < \epsilon \ \forall n \geq n_o$. Using second derivative test, one can easily verify that $\sup_{x \in [0,\infty)} xe^{-nx} = \frac{1}{ne}$. Since $\frac{1}{ne} \to 0$ as $n \to \infty$, $f_n \to f$ uniformly (choose n_o to be any natural number which is greater than $\frac{1}{e\epsilon}$).

It should be noted that pointwise convergence may not imply uniform convergence. For a counter-example, consider the following:

Example 6.1.2. Let $X = [0, 1]$ and let (f_n) be the sequence of functions defined on X by $f_n(x) = x^n$ for all $x \in [0, 1]$. Then each f_n is a continuous function on $([0, 1], | \cdot |)$ and $f_n \to f$ pointwise where $f(x) = 0 \ \forall x \in [0, 1)$ and $f(1) = 1$. Clearly, f is not continuous on $[0, 1]$. This convergence is not uniform: suppose $f_n \to f$ uniformly. Then there exists $n_o \in \mathbb{N}$ such that $|f_n(x) - f(x)| < \frac{1}{2} \ \forall n \geq n_o$ and $\forall x \in [0, 1]$. In particular we have, $x^{n_o} < \frac{1}{2} \ \forall x \in [0, 1)$. This implies that $\lim_{x \to 1-} x^{n_o} \leq \frac{1}{2} \Rightarrow 1 \leq \frac{1}{2}$. A contradiction! Hence the convergence is not uniform. The point 1 in X is a trouble maker. In fact, the reader should check that $f_n \to f$ uniformly on $[0, a]$ for any $a \in (0, 1)$. Shortly, we will see a theoretical justification for such cases.

For a graphic understanding of uniform convergence of a sequence of functions, one can think of the following simple example: let $X = [0, 1]$ and $f_n : [0, 1] \to \mathbb{R}$. Suppose $f_n \to f$ uniformly where $f : [0, 1] \to \mathbb{R}$. Then given $\epsilon > 0$, there exists n_ϵ (depending on ϵ only) such that $|f_n(x) - f(x)| < \epsilon \ \forall n \geq n_\epsilon$ and $\forall x \in [0, 1]$, that is, $f(x) - \epsilon < f_n(x) < f(x) + \epsilon \ \forall n \geq n_\epsilon$ and $\forall x \in [0, 1]$. So if we consider two functions $h_\epsilon = f - \epsilon$ and $g_\epsilon = f + \epsilon$, where $h_\epsilon(x) = f(x) - \epsilon$ and $g_\epsilon(x) = f(x) + \epsilon$ for all $x \in [0, 1]$, then the graphs of

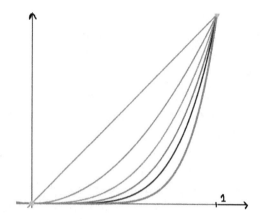

Figure 6.1　Graphs of $f_n(x) = x^n$ on $[0, 1]$ for a few n.

all f_n $(n \geq n_\epsilon)$ lie in between the graphs of h_ϵ and g_ϵ. Informally, we say that the graphs of f_n $(n \geq n_\epsilon)$ lie within the ϵ-band around the graph of f.

In Example 6.1.2, by drawing a few graphs of f_n, one can easily see that the point 1 is creating trouble (Figure 6.1).

Example 6.1.2 clearly shows that the pointwise limit of continuous functions need not be continuous. But the question is: *how large is the set of points of discontinuity of such a pointwise limit function?* The next result says that it is essentially a 'small' set.

Theorem 6.1.1. *Let $f : \mathbb{R} \to \mathbb{R}$ be the pointwise limit of a sequence (f_n) of continuous functions. Then the set A of the points of discontinuity of f can be expressed as a countable union of nowhere dense sets in $\mathbb{R}$.*

Proof. By Corollary 3.7.2,

$$A = \{x \in \mathbb{R} : \omega(f, x) > 0\} = \bigcup_{n \in \mathbb{N}} \left\{ x \in \mathbb{R} : \omega(f, x) \geq \frac{1}{n} \right\}.$$

We need to show that the sets $A_n = \{x \in \mathbb{R} : \omega(f, x) \geq \frac{1}{n}\}$ are nowhere dense in $\mathbb{R}$ for each $n \in \mathbb{N}$. Let us use Proposition 3.6.1. Suppose I is an open interval in $\mathbb{R}$. We will prove the existence of a subinterval $J \subseteq I$ such that $J \cap A_\epsilon = \emptyset$, where $A_\epsilon = \{x \in \mathbb{R} :$

$\omega(f, x) \geq \epsilon\}$ for $\epsilon > 0$. For each $m \in \mathbb{N}$, consider the set

$$B_m = \{x \in I : |f_i(x) - f_j(x)| \leq \epsilon/3 \ \forall \ i, j \geq m\}.$$

Claim. B_m is closed for each $m \in \mathbb{N}$.
For $i, \ j \in \mathbb{N}$, consider the function $h_{i,j} : \mathbb{R} \to \mathbb{R}$ which is defined as

$$h_{i,j}(x) = |f_i(x) - f_j(x)| = |(f_i - f_j)(x)|.$$

Since f_i and f_j are continuous, $f_i - f_j$ is also continuous. Moreover, $x \mapsto |x|$ is also a continuous function. Hence by Corollary 2.1.3, $h_{i,j}$ is continuous. Then Theorem 2.1.1(f) implies that $h_{i,j}^{-1}([0, \epsilon/3])$ is closed in $\mathbb{R}$ for every $i, \ j \in \mathbb{N}$ and consequently, $B_m = \bigcap_{i=m}^{\infty} \bigcap_{j=m}^{\infty} h_{i,j}^{-1}([0, \epsilon/3])$ is closed in $\mathbb{R}$ by Theorem 1.4.2.

Let $B = \bigcup_{m \in \mathbb{N}} B_m$. Then clearly $B \subseteq I$. Since $f_n \to f$ pointwise, the sequence $(f_n(x))$ is Cauchy for each $x \in \mathbb{R}$. Hence for every $x \in I$, there exists $N_x \in \mathbb{N}$ such that

$$|f_i(x) - f_j(x)| \leq \epsilon/3 \ \forall \ i, j \geq N_x.$$

Thus $x \in B_{N_x} \subseteq B \ \forall \ x \in I$. This implies that $I \subseteq B$ and hence $B = I$. We know that $\mathbb{R}$ is a Baire space (Baire category theorem) and $I = B = \bigcup_{m \in \mathbb{N}} B_m$. This means that int $B_{m_o} \neq \emptyset$ for some $m_o \in \mathbb{N}$. Thus, there exists an open interval I_1 which is contained in B_{m_o} and hence contained in I.

Now we claim that $I_1 \cap A_n = \emptyset$. For $x \in I_1$, we have $|f_i(x) - f_j(x)| \leq \epsilon/3 \ \forall \ i, j \geq m_o$. Fixing $i = m_o$ and $j \to \infty$, we get $|f_{m_o}(x) - f(x)| \leq \epsilon/3$. Now consider

$$|f(u) - f(w)| \leq |f(u) - f_{m_o}(u)| + |f_{m_o}(u) - f_{m_o}(w)|$$
$$+ |f_{m_o}(w) - f(w)| \ \ (\star)$$

Further, by uniform continuity of f_{m_o} on any compact subset of I_1 (Theorem 4.3.7), there exists an interval $J \subseteq I_1$ such that $|f_{m_o}(u) - f_{m_o}(w)| < \epsilon/3$ for all $u, \ w \in J$. Then $(\star)$ implies $|f(u) - f(w)| < \epsilon$ for all $u, \ w \in J$. Thus, $\omega(f, x) < \epsilon \ \forall \ x \in J$. Hence $J \cap A_\epsilon = \emptyset$. $\square$

Some of the main applications of uniform convergence are quite evident from the fact that it preserves continuity and Riemann integrability. Recall that the uniform limit f of a sequence (f_n)

of Riemann integrable functions on a closed and bounded interval $[a, b]$ is also Riemann integrable on $[a, b]$. Moreover, the limit and the integral can be interchanged, that is,

$$\int_a^b f = \lim_{n \to \infty} \int_a^b f_n.$$

Now let us prove that the uniform limit of a sequence of continuous functions is always a continuous function.

Theorem 6.1.2. *Let (f_n) be a sequence of real-valued continuous functions on a metric space (X, d). If (f_n) converges uniformly to a real-valued function f on X, then f is continuous on X.*

Proof. Let $a \in X$ and $\epsilon > 0$. We need to show that there exists a $\delta > 0$ such that $|f(x) - f(a)| < \epsilon$ $\forall x \in B(a, \delta)$. Since (f_n) converges uniformly to f on X, there exists $k \in \mathbb{N}$ such that $|f_k(x) - f(x)| < \epsilon/3$ $\forall x \in X$ $(\star)$

Since f_k is continuous on X, f_k is continuous at a. Hence there exists a $\delta > 0$ such that

$$|f_k(x) - f_k(a)| < \epsilon/3 \quad \forall x \in B(a, \delta) \quad (\star\star)$$

Now

$$|f(x) - f(a)| = |f(x) - f_k(x) + f_k(x) - f_k(a) + f_k(a) - f(a)|$$
$$\leq |f(x) - f_k(x)| + |f_k(x) - f_k(a)| + |f_k(a) - f(a)|$$
$$< \epsilon/3 + \epsilon/3 + \epsilon/3 \quad \forall x \in B(a, \delta) \text{ (by } (\star) \text{ and } (\star\star)).$$

This implies that $|f(x) - f(a)| < \epsilon$ $\forall x \in B(a, \delta)$. Consequently, f is continuous at a. But a was chosen arbitrarily from X. Hence f is continuous at each point of X, that is, f is continuous on X. $\square$

Corollary 6.1.1. *The space $C[a, b]$ equipped with the uniform metric D is a closed subset of $(B[a, b], D)$.*

Though Theorem 6.1.2 cannot be used for proving uniform convergence of a sequence but it could be used for disproving the uniform convergence of sequences of continuous functions. For example, consider the sequence of functions (f_n) where f_n is defined on $[0, \infty)$ as: $f_n(x) = 0$ for $x \geq 1/n$ and $f_n(x) = -nx + 1$ for $0 \leq x \leq 1/n$. See Figure 6.2.

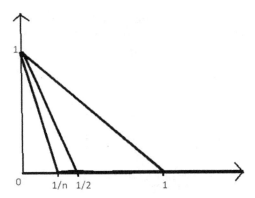

Figure 6.2 Graphs of f_1, f_2 and f_n.

Then (f_n) is a sequence of continuous functions which converges pointwise to the function f, where $f(0) = 1$ and $f(x) = 0$ for $x \in (0, \infty)$. By Theorem 6.1.2, the convergence is not uniform as f is not continuous. In fact, Theorem 6.1.2 can be conveniently used to show that the sequence of functions defined in Example 6.1.2 does not converge uniformly.

Remarks. (a) If (f_n) is a sequence of uniformly continuous functions from (X, d) to $\mathbb{R}$ which converges uniformly to $f : (X, d) \to \mathbb{R}$, then f is also uniformly continuous (proof is similar to that of Theorem 6.1.2).

(b) The uniform limit of a sequence of Cauchy-continuous functions is also Cauchy-continuous.

But one should note that the uniform limit of a sequence of Lipschitz functions need not be Lipschitz.

Example 6.1.3. Let $f_n(x) = \sqrt{x + \frac{1}{n}}$ for $x \in [0, 1]$. Then using Proposition 2.4.1, it can be verified that f_n is Lipschitz for all $n \in \mathbb{N}$. Further, $f_n \to \sqrt{x}$ uniformly (check !). But the function $f(x) = \sqrt{x}$ is not Lipschitz on $[0, 1]$ because

$$\frac{|\sqrt{x} - 0|}{|x - 0|} \to \infty \text{ as } x \to 0^+.$$

Theorem 6.1.3. *Let (f_n) be a sequence of real-valued Lipschitz functions on a metric space (X, d) with Lipschitz constant $M_n > 0$. If*

$f_n \to f$ *pointwise and* $\sup_{n \in \mathbb{N}} M_n = M \in \mathbb{R}$, *then* f *is Lipschitz with Lipschitz constant* M. *Moreover, the convergence of* (f_n) *is uniform on every totally bounded subset of* X.

Proof. Since each f_n is Lipschitz,

$$|f_n(x) - f_n(y)| \leq M_n d(x,y) \leq M d(x,y) \ \forall \ x, \ y \in X.$$

By Proposition 2.5.1,

$$|f(x) - f(y)| = \lim_{n \to \infty} |f_n(x) - f_n(y)| \leq M d(x,y),$$

for all $x, \ y \in X$. Let A be a totally bounded subset of X and let $\epsilon > 0$. Then there exist $x_1, \ldots, x_k \in X$ such that $A \subseteq \bigcup_{i=1}^{k} B(x_i, \epsilon)$. Since f is the pointwise limit of (f_n), there exists $n_o \in \mathbb{N}$ such that $|f_n(x_i) - f(x_i)| \leq \epsilon$ for all $n \geq n_o$ and for all $i \in \{1, 2, \ldots, k\}$. Let $a \in A$. Then $d(a, x_m) < \epsilon$ for some $m \in \{1, 2, \ldots, k\}$. Now

$$|f_n(a) - f(a)| \leq |f_n(a) - f_n(x_m)| + |f_n(x_m) - f(x_m)|$$
$$+ |f(x_m) - f(a)|$$
$$\leq M d(a, x_m) + \epsilon + M d(x_m, a) < (2M + 1)\epsilon$$

for all $n \geq n_o$. Since a was an arbitrary element of A, $f_n \to f$ uniformly on A. $\qquad\square$

If (f_n) and (g_n) are sequences of real-valued functions which converge uniformly on X, then it is easy to see that the sequence $(\alpha f_n + \beta g_n)$ also converges uniformly on X, where $\alpha, \ \beta \in \mathbb{R}$. *What can you say about the pointwise product of the two sequences?* Clearly, the pointwise product $(f_n g_n)$ of the two sequences converges pointwise but in general it is not uniformly convergent as seen in the following example.

Example 6.1.4. Let $f_n : [0, \infty) \to \mathbb{R}$ be defined as $f_n(x) = x + \frac{1}{n}$. Then the sequence (f_n) converges uniformly to the identity function f. Furthermore, the sequence (f_n^2) converges pointwise to f^2. Suppose, if possible, the convergence is uniform. Then there exists $n_o \in \mathbb{N}$ such that $|f_n^2(x) - f^2(x)| < 1 \ \forall \ n \geq n_o$ and $\forall \ x \geq 0$. In particular, $|f_n^2(n^2) - f^2(n^2)| = 2n + \frac{1}{n^2} < 1$ for all $n \geq n_o$. This gives a contradiction.

In the previous example, note that the pointwise limit function for the sequence is unbounded. Now in the next proposition it is proved that the pointwise product of uniformly convergent sequences is uniformly convergent provided the limit function is bounded.

Proposition 6.1.1. *Let f_n and g_n be real-valued functions defined on a set X. Suppose the sequences (f_n) and (g_n) converge uniformly to the bounded functions f and g, respectively. Then the pointwise product of the two sequences $(f_n g_n)$ converges uniformly to fg, where $(f_n g_n)(x) = f_n(x)g_n(x)$.*

Proof. Since f and g are bounded, there exists $M > 0$ such that $|f(x)| < M$ and $|g(x)| < M$ for all $x \in X$. Since (f_n) is uniformly convergent to f, there exists $n_o \in \mathbb{N}$ such that $|f_n(x) - f(x)| < 1$ $\forall\, n \geq n_o$ and $\forall\, x \in X$. Consequently,

$$|f_n(x)| \leq |f_n(x) - f(x)| + |f(x)| < 1 + M$$

$\forall\, n \geq n_o$ and $\forall\, x \in X$. Let $\epsilon > 0$. The uniform convergence of (f_n) and (g_n) implies that there exists $n_1 \in \mathbb{N}$ such that

$$|f_n(x) - f(x)| < \frac{\epsilon}{2M} \quad \text{and} \quad |g_n(x) - g(x)| < \frac{\epsilon}{2(M+1)}$$

$\forall\, n \geq n_1$ and $\forall\, x \in X$. Hence for $n \geq \max\{n_o, n_1\}$ and $x \in X$, we have

$$\begin{aligned}
|f_n(x)g_n(x) - f(x)g(x)| &= |f_n(x)g_n(x) - f_n(x)g(x) + f_n(x)g(x) \\
&\quad - f(x)g(x)| \leq |f_n(x)||g_n(x) - g(x)| \\
&\quad + |g(x)||f_n(x) - f(x)| \\
&< (M+1)\frac{\epsilon}{2(M+1)} + M\frac{\epsilon}{2M} = \epsilon.
\end{aligned}$$

This means that $(f_n g_n)$ converges uniformly to fg. $\qquad\square$

Exercises

6.1.1. Let (f_n) be a sequence of real-valued functions defined on a non-empty subset X of $\mathbb{R}$. Suppose f_n is increasing for each $n \in \mathbb{N}$, that is, $x < y \Rightarrow f_n(x) \leq f_n(y)$. Then show that the pointwise limit (if exists) of (f_n) is also an increasing function. Does the result still hold if we replace increasing with strictly increasing?

6.1.2. Let (f_n) be a sequence of real-valued continuous functions on (X, d). If (f_n) converges uniformly to a real-valued function f on every compact subset of X, then show that f is continuous on X.

6.1.3. With the help of an example, show that the uniform limit of a sequence of discontinuous functions need not be discontinuous.

6.1.4. Show that the function $f(x) = x^2$, defined on $\mathbb{R}$, is not the uniform limit of any sequence of Lipschitz functions.

6.1.5. Show that the normed linear space $(\text{Lip}(X), \|\cdot\|)$ is complete, where

$$\|f\| = \max\{\ |f(x_o)|,\ L(f)\}\ \text{ for } f \in \text{Lip}(X),$$

for a fixed $x_o \in X$. (Refer to Exercise 2.4.11.)

6.1.6. Show that the function $f : (X, d) \to (Y, \rho)$ is uniformly continuous if and only if the sequence of functions $\langle \omega_n(f, \cdot) \rangle$ converges uniformly to the zero function on X. For definition of the function $\omega_n(f, \cdot) : (X, d) \to [0, \infty]$, see Section 3.7.

6.1.7. Using Theorem 6.1.2, construct more examples of sequences of functions which are pointwise convergent but not uniformly convergent.

6.2 Dini's Theorem

Recall that given a non-empty set X, $\mathbb{R}^X$ denotes the set of all functions from X to $\mathbb{R}$. In fact, it is a vector space over $\mathbb{R}$ under the usual addition of functions on X and the usual scalar multiplication of f in $\mathbb{R}^X$ by a scalar $\alpha \in \mathbb{R}$. Moreover, with respect to these operations, the collection $B(X)$ of all bounded functions in $\mathbb{R}^X$ is also a vector space over $\mathbb{R}$. In Example 3.1.6, it was shown that the metric space $(B(X), D)$ is complete where D is the uniform metric. Now in order to talk about another special subset of $\mathbb{R}^X$, we consider a metric d on the set X. Let $C(X)$ denotes the set of all real-valued continuous functions on the metric space (X, d). It can be easily checked that $C(X)$ is a vector space over $\mathbb{R}$. Thus $C(X)$ and $B(X)$

are both linear subspaces of $\mathbb{R}^X$. Consequently, $C(X) \cap B(X)$ is also a linear subspace of $\mathbb{R}^X$. Note that $C(X) \cap B(X) = \{f : X \to \mathbb{R} : f$ is continuous as well as bounded$\}$. Let us denote $C(X) \cap B(X)$ by $C_b(X)$ which is the set of all bounded real-valued continuous functions on (X, d). Since $C_b(X) \subseteq B(X)$, we can consider the metric induced by the uniform metric D on $C_b(X)$. Here note that a sequence (f_n) in $B(X)$ converges to some f in $B(X)$ if and only if $D(f_n, f) \to 0$ as $n \to \infty$ if and only if (f_n) converges uniformly to f. Due to this reason, the metric D is referred to as the 'uniform' metric. Similarly, a sequence which is Cauchy in $(B(X), D)$ is known as *uniformly Cauchy*. Let us define it precisely for a sequence in $\mathbb{R}^X$.

Definition 6.2.1. A sequence (f_n) of real-valued functions defined on a non-empty set X is said to be **uniformly Cauchy** if for every $\epsilon > 0$, there exists $n_o \in \mathbb{N}$ such that

$$|f_n(x) - f_m(x)| < \epsilon \ \forall \ n, \ m \geq n_o \text{ and } \forall \ x \in X.$$

It is easy to see that every uniformly convergent sequence of real-valued functions on a set X is uniformly Cauchy. In fact, one can easily imitate the proof given in Example 3.1.6 to prove that the converse is also true (note that in the proof we are using the completeness of $\mathbb{R}$). Hence the class of uniformly Cauchy sequences is same as that of uniformly convergent sequences. Let us state it precisely.

Proposition 6.2.1. *Let (f_n) be a sequence of real-valued functions defined on a non-empty set X. Then the sequence (f_n) is uniformly convergent if and only if it is uniformly Cauchy.*

Remark. If (f_n) is a sequence of functions from a set X to a complete metric space (Y, ρ), then (f_n) is uniformly convergent if and only if it is uniformly Cauchy.

Now we show that the metric subspace $(C_b(X), D)$ of $(B(X), D)$ is also complete.

Theorem 6.2.1. *If (X, d) is a metric space, then $(C_b(X), D)$ is a complete metric space.*

Proof. Let (f_n) be a Cauchy sequence in $(C_b(X), D)$. Hence (f_n) is Cauchy in the complete metric space $(B(X), D)$. Then there exists $f \in B(X)$ such that $D(f_n, f) \to 0$ as $n \to \infty$. By the definition of the metric D, (f_n) converges uniformly to f. But then by Theorem 6.1.2, f is continuous on X, that is, $f \in C(X)$. Hence $f \in B(X) \cap C(X) = C_b(X)$. This implies that (f_n) converges to f in $(C_b(X), D)$. So the metric space $(C_b(X), D)$ is complete. $\qquad\square$

Corollary 6.2.1. *If (X, d) is a compact metric space, then the metric space $(C(X), D)$ is complete.*

Proof. Since (X, d) is compact, every real-valued continuous function on (X, d) is bounded (Theorem 4.3.4). Hence $C(X) = C_b(X)$. Now the conclusion follows from Theorem 6.2.1. $\qquad\square$

Since pointwise convergent sequences need not be uniformly convergent in general scenario and uniformly convergent sequences play a vital role especially in real analysis, our next task is to analyze the conditions under which pointwise convergence implies uniform convergence. In this regard, we need the following definitions.

Definitions 6.2.2. A sequence (f_n) of real-valued functions on a set X is said to be **increasing** if $f_n \leq f_{n+1}$ holds for all $n \in \mathbb{N}$, that is, $f_n(x) \leq f_{n+1}(x)$ holds for all $n \in \mathbb{N}$ and for all $x \in X$.

Similarly, (f_n) is called **decreasing** if $f_n \geq f_{n+1}$ holds for all $n \in \mathbb{N}$, that is, $f_n(x) \geq f_{n+1}(x)$ holds for all $n \in \mathbb{N}$ and for all $x \in X$.

An increasing or a decreasing sequence of functions is referred to as a **monotone** sequence of functions.

For illustration, think about the sequence (f_n) where $f_n(x) = x/n$ for $n \in \mathbb{N}$ and $x \in [0, \infty)$. Clearly, (f_n) is a decreasing sequence. Monotone sequence of functions plays a big role in measure theory. In fact, now we are going to exhibit their role in uniform convergence of sequence of functions.

Theorem 6.2.2 (Dini's theorem). *Let (X, d) be a compact metric space. If a monotone sequence in $C(X)$ converges pointwise to a continuous function, then the convergence is uniform.*

Proof. Suppose that the sequence (f_n) in $C(X)$ is increasing and converges pointwise to some $f \in C(X)$, that is, $f_n(x) \uparrow f(x)$ holds for each $x \in X$. This implies that for each $x \in X$, $(f_n(x))$ is an increasing sequence in $\mathbb{R}$ which converges to a real number $f(x)$. So $f(x) = \sup_n f_n(x)$. Now let $\epsilon > 0$. For each $n \in \mathbb{N}$, define

$$U_n = \{x \in X : f(x) - f_n(x) < \epsilon\} = (f - f_n)^{-1}(-\epsilon, \epsilon).$$

Note that $f(x) - f_n(x) \geq 0 \; \forall \; n \in \mathbb{N}$ and $\forall \; x \in X$. Hence $(f - f_n)^{-1}([0, \epsilon)) = (f - f_n)^{-1}(-\epsilon, \epsilon)$. Since f and f_n are continuous, $f - f_n$ is also continuous. Hence $(f - f_n)^{-1}(-\epsilon, \epsilon)$ is open in (X, d), that is, each U_n is open in (X, d). Since $f_n(x) \leq f_{n+1}(x) \; \forall \; x \in X$, $U_n \subseteq U_{n+1}$ ($\star$).

Moreover, since $f_n(x) \uparrow f(x)$ holds for each $x \in X$, given $x \in X$, there exists $n_x \in \mathbb{N}$ such that $f(x) - f_{n_x}(x) < \epsilon$, that is, $x \in U_{n_x}$. Hence $X = \bigcup_{n=1}^{\infty} U_n$. Now (X, d) is compact and $\{U_n : n \in \mathbb{N}\}$ is an open cover of X. Thus, there exist $k_1, \ldots, k_l$ in $\mathbb{N}$ such that $X = \bigcup_{i=1}^{l} U_{k_i}$. Let $m = \max\{k_1, \ldots, k_l\}$. Then $X = U_m = U_n \; \forall \; n \geq m$ by ($\star$). This implies that $0 \leq f(x) - f_n(x) < \epsilon \; \forall \; n \geq m$ and $\forall \; x \in X$. But this precisely says that (f_n) converges uniformly to f on X.

Note that in case (f_n) is a decreasing sequence, then we can work with the sequence $(-f_n)$. $\qquad \square$

Remark. The reader should carefully observe the requirement of each and every condition given in the hypothesis of Dini's theorem. If we drop any one of these conditions, then the conclusion may not hold true as seen in the following examples:

(a) *Compactness of (X, d):* let $X = [0, \infty)$ and $f_n : [0, \infty) \to \mathbb{R}$ be defined as $f_n(x) = x/n$ for all $n \in \mathbb{N}$. Then the sequence (f_n) is a monotone sequence of continuous functions which converges pointwise to the zero function f. Suppose $f_n \to f$ uniformly. Then there exists $n_o \in \mathbb{N}$ such that $\frac{x}{n} < \frac{1}{2} \; \forall \; n \geq n_o$ and $\forall \; x \in X$. In particular, it should hold for $x = n_o$ and $n = n_o$. Then we get a contradiction. Thus the convergence is not uniform.

(b) *Monotonicity of (f_n):* let $X = \{0, \; 1/n : n \in \mathbb{N}\}$ and for $n \in \mathbb{N}$ define $f_n : X \to \mathbb{R}$ as $f_n(1/n) = 1$ and zero otherwise. Then it can be easily verified that the sequence (f_n) of continuous functions converges pointwise to the zero function f but the convergence is not uniform (particularly, take $x = 1/n_o$ and $n = n_o$).

(c) *Continuity of f_n's:* let $X = [0, 1]$ and for $n \in \mathbb{N}$ define $f_n : X \to \mathbb{R}$ as $f_n(x) = 1$ for $x \in (0, \frac{1}{n}]$ and zero otherwise. Then the monotone sequence (f_n) converges pointwise to the zero function

f but the convergence is not uniform (particularly, take $x = 1/n_o$ and $n = n_o$).

Interestingly, the converse of the Dini's theorem also holds true as seen in the next result.

Theorem 6.2.3. *For a metric space (X, d), the following statements are equivalent:*

(a) *X is compact.*
(b) *Every increasing sequence (f_n) of functions in $C(X)$ converging pointwise to a function f in $C(X)$, converges uniformly.*

Proof. We only need to prove (b) $\Rightarrow$ (a). For that we use Theorem 4.3.4.

Let $f : X \to \mathbb{R}$ be a continuous function. Without loss of generality, assume that $f \geq 0$ (otherwise work with $|f|$). Now for each $n \in \mathbb{N}$, define $f_n(x) = \min\{f(x), n\}$ for $x \in X$. Note that $f_n \in C(X)$ (see Exercise 2.2.2) and $(f_n(x))$ increases to $f(x)$ for each $x \in X$. Then by (b), $f_n \to f$ uniformly. Hence $\exists \, n_o \in \mathbb{N}$ such that

$$0 \leq f(x) - f_n(x) < 1 \ \ \forall \, n \geq n_o \quad \text{and} \quad \forall \, x \in X.$$

Consequently, $0 \leq f(x) < 1 + f_{n_o}(x) \leq 1 + n_o \ \ \forall \, x \in X$. Thus, f is bounded.

$\square$

Exercises

6.2.1. Let $f_n : \mathbb{R} \to \mathbb{R}$ be continuous for every $n \in \mathbb{N}$. If (f_n) is uniformly Cauchy on $\mathbb{Q}$, then show that it is uniformly Cauchy on $\mathbb{R}$.

6.2.2. Let (f_n) be a sequence of non-negative continuous functions defined on $[a, b]$. Suppose that the series $\sum_{n \in \mathbb{N}} f_n$ converges pointwise to a continuous function on $[a, b]$. Show that the convergence is uniform on $[a, b]$. Note that a series $\sum_{n \in \mathbb{N}} f_n$ of functions in $\mathbb{R}^X$ is said to be pointwise (uniformly) convergent to $f \in \mathbb{R}^X$ if the corresponding sequence of partial sums (s_n), where $s_n(x) = f_1(x) + \ldots + f_n(x)$, is pointwise (uniformly) convergent to f.

6.2.3. Let $0 < a < 1$ be a fixed number. Prove that the sequence of functions (f_n), where $f_n : [0, a] \to \mathbb{R}$ is defined as $f_n(x) = x^n$ for all $n \in \mathbb{N}$, converges uniformly.

6.2.4. Suppose that (f_n) is a sequence of continuous functions from a metric space (X, d) to $\mathbb{R}$. Show that if $f_n \to f$ uniformly on a dense subset D of X, then $f_n \to f$ uniformly on X. Does the result still hold if we drop the continuity of the functions?

6.2.5 (Dini's theorem for upper semi-continuous functions). If (f_n) is a decreasing sequence of upper semi-continuous real-valued functions on a compact metric space (X, d) that converges pointwise to a lower semi-continuous function, then prove that the convergence is uniform.

6.3   Equicontinuity

When we talk about continuity of a function at a point x, then the positive number δ which we choose depends on both ϵ and x (see Definition 2.1.1). Now if we consider a collection of continuous functions then we will get a stronger notion if some δ can be chosen which depends on ϵ and x but not on the function in the collection. This section is devoted to the study of this notion. Interestingly, this will help in characterizing the compact subsets of $(C(X), D)$, where (X, d) is a compact metric space and D is the supremum (uniform) metric on $C(X)$.

So let us start the discussion.

Definition 6.3.1. Let (X, d) be a metric space and $\emptyset \neq F \subseteq C(X)$. For a given $x \in X$, F is said to be **equicontinuous** at x if given $\epsilon > 0$, there exists a $\delta > 0$ such that $|f(x) - f(y)| < \epsilon$ whenever $y \in B(x, \delta)$ and $f \in F$ (same δ works for all f in F).

We say that F is equicontinuous on X or simply F is equicontinuous if F is equicontinuous at each $x \in X$.

Example 6.3.1. Let $X = [0, 1]$ and $F = \{f_n : n \in \mathbb{N}\} \subseteq C(X)$, where $f_n(x) = x^n$. Then the collection F is not equicontinuous at $x = 1$: given any t with $0 < t < 1$, then we know that $t^n \to 0$ as $n \to \infty$. Thus we have $|f_n(1) - f_n(t)| = 1 - t^n > \frac{1}{2}$ for all sufficiently large n, and so the condition for equicontinuity of F fails at 1 (for $\epsilon = \frac{1}{2}$).

However, F is equicontinuous on $[0, 1)$. Given $0 \leq x < 1$ and $\epsilon > 0$, choose a with $x < a < 1$. Since the series $\sum_{n=1}^{\infty} na^{n-1}$ converges (apply Ratio Test), there exists $b > 0$ such that $na^{n-1} \leq b \; \forall \; n \in \mathbb{N}$. Choose $\delta = \min\{a - x, \frac{\epsilon}{b}\}$. Now for a non-negative number t such that $t \in (x - \delta, x + \delta)$ and for all $n \in \mathbb{N}$, we have

$$|f_n(t) - f_n(x)| = |t^n - x^n| = \left| (t - x) \sum_{k=1}^{n} t^{n-k} x^{k-1} \right|$$

$$\leq |t - x| n a^{n-1} \text{ because } t < x + \delta \leq a; \; x < a$$

$$< \left(\frac{\epsilon}{b} \right) . b = \epsilon.$$

Hence f is equicontinuous at x. Since x was chosen arbitrarily from $[0, 1)$, f is equicontinuous on $[0, 1)$. Moreover, the sequence (f_n) converges uniformly on $[0, a]$ whenever $0 < a < 1$, but it does not converge uniformly on $[0, 1]$ (see Example 6.1.2). In fact, (f_n) converges uniformly on $K \subseteq [0, 1]$ if and only if 1 is not an accumulation point of K (verify!).

Our next result gives sufficient conditions to deduce uniform convergence on all compact subsets from mere pointwise convergence on a dense subset.

Theorem 6.3.1. *Let (X, d) be a metric space and D be a dense subset of X. Suppose (f_n) is an equicontinuous sequence in $C(X)$ such that for each $x \in D$, $(f_n(x))$ converges in $\mathbb{R}$. Then there is a function f in $C(X)$ such that (f_n) converges uniformly to f on every compact subset K of X.*

Proof. Let $x \in X$ and $\epsilon > 0$ be given. By equicontinuity of (f_n), there is a $\delta_x > 0$ such that

$$|f_n(t) - f_n(x)| < \frac{\epsilon}{3} \; \forall \; t \in B(x, \delta_x) \text{ and } \forall \; n \in \mathbb{N} \qquad (1)$$

Since $\overline{D} = X$, $\exists \; t_o \in B(x, \delta_x) \cap D$. Since the sequence $(f_n(t_o))$ converges in $\mathbb{R}$, it is a Cauchy sequence. Hence $\exists \; n_o \in \mathbb{N}$ such that

$$|f_m(t_o) - f_n(t_o)| < \frac{\epsilon}{3} \; \forall \; m, \; n \geq n_o \qquad (2)$$

By (1) and (2), we get

$$|f_m(x) - f_n(x)| \leq |f_m(x) - f_m(t_o)| + |f_m(t_o) - f_n(t_o)|$$
$$+ |f_n(t_o) - f_n(x)| < \epsilon \ \forall \ m, \ n \geq n_o.$$

This shows that $(f_n(x))$ is a Cauchy sequence in $(\mathbb{R}, |\cdot|)$. Since $(\mathbb{R}, |\cdot|)$ is complete, $(f_n(x))$ converges in $\mathbb{R}$. Hence, we obtain a function $f : X \to \mathbb{R}$ such that (f_n) converges pointwise to f on X, that is, for each $x \in X$, $\lim_{n\to\infty} f_n(x) = f(x)$.

Now we show that f is continuous on X. Let $x \in X$ and $\epsilon > 0$ be given. Choose δ_x as before, so that (1) is obtained. For any given $t \in B(x, \delta_x)$, choose n sufficiently large so that we have

$$|f(t) - f_n(t)| < \frac{\epsilon}{3} \quad \text{and} \quad |f_n(x) - f(x)| < \frac{\epsilon}{3}.$$

Then by using (1), we get

$$|f(t) - f(x)| \leq |f(t) - f_n(t)| + |f_n(t) - f_n(x)| + |f_n(x) - f(x)|$$
$$< \epsilon \ \forall \ t \in B(x, \delta_x). \tag{3}$$

Hence f is continuous at x.

Now finally, we show that $f_n \to f$ uniformly on every compact subset K of X. Let $\epsilon > 0$. For each $x \in K$, choose a $\delta_x > 0$ such that (1) is obtained. Then by reasoning as in the preceding paragraph, we obtain (3). Since K is compact in X, there exists a finite subset $\{x_1, \ldots, x_n\}$ of K such that $K \subseteq \bigcup_{j=1}^{n} B(x_j, \delta_{x_j})$. Now choose m sufficiently large such that

$$|f(x_j) - f_n(x_j)| < \frac{\epsilon}{2} \ \forall \ n \geq m \text{ and for } j \in \{1, \ldots, n\}. \tag{4}$$

Now given $t \in K$, choose j $(1 \leq j \leq n)$ such that $t \in B(x_j, \delta_{x_j})$. By using (3), (4) and (1) we have

$$|f(t) - f_n(t)| \leq |f(t) - f(x_j)| + |f(x_j) - f_n(x_j)| + |f_n(x_j) - f_n(t)|$$
$$< \epsilon + \frac{\epsilon}{2} + \frac{\epsilon}{3} \ \forall \ n \geq m < 2\epsilon \ \forall \ n \geq m.$$

Hence $|f(t) - f_n(t)| < 2\epsilon \ \forall \, t \in K$ and $\forall \, n \geq m$. But this precisely means that (f_n) converges uniformly to f on K. $\qquad\square$

Remark. Note that equicontinuity of the sequence (f_n) cannot be dropped in the previous theorem: let $X = \{0, \ 1/n : n \in \mathbb{N}\}$ and for $n \in \mathbb{N}$ define $f_n : X \to \mathbb{R}$ as $f_n(1/n) = 1$ and zero otherwise. The collection $\{f_n : n \in \mathbb{N}\}$ is not equicontinuous at 0. Moreover, the sequence (f_n) of continuous functions converges pointwise to the zero function f but the convergence is not uniform on the compact set X (particularly, take $x = 1/n_o$ and $n = n_o$).

Interestingly, the previous result can be thought of as 'equicontinuous' version of Dini's theorem. Let us state it precisely.

Corollary 6.3.1. *Let (X, d) be a compact metric space. If (f_n) is an equicontinuous sequence in $C(X)$ which converges pointwise to some real-valued function f on X, then the convergence is uniform.*

The proof of our next result is based on the classical diagonal sequence argument. This result is often useful in proving powerful theorems.

Theorem 6.3.2. *Let $D = \{x_1, x_2, \ldots\}$ be a non-empty countable set. For each n, let $f_n : D \to \mathbb{R}$ be a function such that for each $x \in D$, the set $\{f_n(x) : n \in \mathbb{N}\}$ is bounded in $\mathbb{R}$. Then there exists a subsequence (f_{n_k}) of (f_n) such that $(f_{n_k}(x))$ converges in $\mathbb{R}$ for each $x \in D$.*

Proof. Note that for each $x \in D$, $\overline{\{f_n(x) : n \in \mathbb{N}\}}$ is compact in $(\mathbb{R}, |\cdot|)$ by Heine–Borel Theorem. Hence for each $x \in D$, $(f_n(x))$ has a convergent subsequence in $(\mathbb{R}, |\cdot|)$. Consequently, for $x = x_1$, there exists a subsequence (g_n^1) of (f_n) such that $\lim\limits_{n \to \infty} g_n^1(x_1)$ exists in $\mathbb{R}$. Similarly for $x = x_2$, there exists a subsequence (g_n^2) of (g_n^1) such that $\lim\limits_{n \to \infty} g_n^2(x_2)$ exists in $\mathbb{R}$. By continuing in this way, we can choose (inductively) sequences (g_n^i), $i = 1, 2, \ldots$, such that

(a) (g_n^1) is a subsequence of (f_n),
(b) (g_n^{i+1}) is a subsequence of (g_n^i) for each $i = 1, 2, \ldots$ and
(c) $\lim\limits_{n \to \infty} g_n^i(x_i)$ exists in $\mathbb{R}$ for each $i = 1, 2, \ldots$

Now consider the diagonal sequence (h_n) where $h_n = g_n^n$ for each n. Note that (h_n) is a subsequence of (f_n) such that $\lim_{n\to\infty} h_n(x_i)$ exists in $\mathbb{R}$ for each i. Then this sequence (h_n) is our desired subsequence. $\square$

Now we are ready to prove a version of Ascoli–Arzelà Theorem.

Theorem 6.3.3. *Let (X, d) be a separable metric space. Let F be a non-empty equicontinuous subset of $C(X)$ such that for each $x \in X$, the set $\{f(x) : f \in F\}$ is bounded in $\mathbb{R}$. Then each sequence (f_n) in F has a subsequence that converges pointwise to a function f in $C(X)$. Moreover, this convergence is uniform on every compact subset of X.*

Proof. Let D be a countable dense subset of X and let (f_n) be a sequence in F. By Theorem 6.3.2, there exists a subsequence (f_{n_k}) of (f_n) such that $(f_{n_k}(x))$ converges in $\mathbb{R}$ for each $x \in D$. Now apply Theorem 6.3.1 to this subsequence to obtain the desired function f. $\square$

Before moving on to the main Ascoli–Arzelà Theorem, we would like to formalize one of the hypothesis of Theorems 6.3.2 and 6.3.3 in the following definition.

Definitions 6.3.2. Let S be a non-empty collection in $\mathbb{R}^X$, where X is a non-empty set. Then S is called

(a) **Pointwise bounded** if, for each $x \in X$, $\{f(x) : f \in S\}$ is bounded in $\mathbb{R}$.
(b) **Uniformly bounded** if there exists $M > 0$ such that $|f(x)| \leq M \ \forall \, x \in X$ and $\forall \, f \in S$.

Remarks. (a) By Heine–Borel Theorem, $\{f(x) : f \in S\}$ is bounded in $\mathbb{R}$ if and only if its closure in $\mathbb{R}$, $\overline{\{f(x) : f \in S\}}$, is compact.
(b) If S is a non-empty subset of $C_b(X) = \{f \in C(X) : f(X) \text{ is bounded in } \mathbb{R}\}$ then S is uniformly bounded if and only if S is bounded in the normed linear space $(C_b(X), \| \cdot \|_\infty)$, that is, $\exists \, M > 0$ such that $\|f\|_\infty = \sup_{x \in X} |f(x)| \leq M \ \forall \, f \in S$. Note that the supremum metric D on $C_b(X)$ is the metric induced by the norm $\| \cdot \|_\infty$ on $C_b(X)$.

Evidently, every uniformly bounded collection of functions is pointwise bounded but the converse is not true in general. For example, for $n \in \mathbb{N}$, define $f_n : \mathbb{R} \to \mathbb{R}$ as $f_n(n) = n$ and else it is zero. Then the collection $\{f_n : n \in \mathbb{N}\}$ is pointwise bounded but not uniformly bounded.

Recall that Heine–Borel Theorem exhibited the class of compact subsets in $\mathbb{R}$ in a precise manner. Now, let us talk about the precise collection of compact subsets of $(C(X), D)$, where (X, d) is compact. Since compactness is equivalent to completeness and total boundedness (Theorem 4.2.6), we first characterize the totally bounded subsets of $(C(X), D)$.

Theorem 6.3.4 (Ascoli–Arzelà Theorem). *Let (X, d) be a compact metric space and let S be a non-empty subset of $C(X)$. Then the following statements are equivalent:*

(a) *S is totally bounded in $(C(X), D)$,*
(b) *S is uniformly bounded and equicontinuous and*
(c) *S is pointwise bounded and equicontinuous.*

Proof. (a) $\Rightarrow$ (b): Since S is totally bounded, it is bounded in $(C(X), D)$. Hence there exists $M > 0$ such that $D(f, f_o) < M \ \forall \ f \in S$, where f_o is the zero function in $C(X)$. Therefore, $\sup_{x \in X} |f(x)| < M \ \forall \ f \in S$. This implies that S is uniformly bounded.

To prove that S is equicontinuous, let $\varepsilon > 0$ and $x \in X$. Since S is totally bounded, there exist $f_1, \ldots, f_n \in S$ such that $S \subseteq \bigcup_{i=1}^{n} B(f_i, \varepsilon)$, where $B(f_i, \varepsilon) = \{g \in C(X) : D(f_i, g) < \varepsilon\}$. Since $S \subseteq C(X)$, f_i is continuous for all $i \in \{1, 2, \ldots, n\}$. Consequently, we can choose a $\delta_x > 0$ such that $|f_i(y) - f_i(x)| < \varepsilon$ for all $y \in B(x, \delta_x)$ and for all $i = 1, \ldots, n$.

Now let $y \in B(x, \delta_x)$ and $f \in S$. Then $f \in B(f_i, \varepsilon)$ for some $i \in \{1, 2, \ldots, n\}$. So $|f(y) - f_i(y)| < \varepsilon$ and $|f(x) - f_i(x)| < \varepsilon$. Thus

$$|f(y) - f(x)| \leq |f(y) - f_i(y)| + |f_i(y) - f_i(x)| + |f_i(x) - f(x)|$$

$$< \varepsilon + \varepsilon + \varepsilon = 3\varepsilon.$$

This shows that S is equicontinuous at x. But x was chosen arbitrarily. Hence S is equicontinuous on X.

(b) $\Rightarrow$ (c): This is immediate.

(c) $\Rightarrow$ (a): We show that every sequence (f_n) in S has a convergent subsequence in $(C(X), D)$. First note that since (X, d) is a compact metric space, it is separable (Corollary 4.2.2). Hence by

Theorem 6.3.3, (f_n) has a subsequence (f_{n_k}) that converges point-wise on X to a function $f \in C(X)$. Moreover, (f_{n_k}) converges uniformly to f on every compact subset of X. Since (X, d) is itself compact, (f_{n_k}) converges uniformly to f on X. This precisely means that (f_{n_k}) converges to f in $(C(X), D)$. Hence S is totally bounded by Theorem 3.2.3. $\qquad\square$

Remark. The reader should carefully observe the requirement of each and every condition given in Ascoli–Arzelà Theorem. If we drop any one of these conditions, then the conclusion may not hold true as seen in the following examples:

(i) *Compactness of (X, d):* the sequence (f_n) in $(C_b(\mathbb{R}), D)$ defined by $f_n(x) = e^{-(x-n)^2}$ is a pointwise bounded equicontinuous family (verify!). But $D(f_n, f_m) \geq 1 - \frac{1}{e}$ for $n \neq m$. So $\{f_n : n \in \mathbb{N}\}$ is not totally bounded.

(ii) *Equicontinuity of S:* let $X = \{0, \frac{1}{n} : n \in \mathbb{N}\}$ and for $n \in \mathbb{N}$ define $f_n : X \to \mathbb{R}$ as $f_n(x) = 1$ for $x \leq \frac{1}{n}$ and zero otherwise. Then the collection $S = \{f_n : n \in \mathbb{N}\}$ is uniformly bounded. But $D(f_n, f_m) = 1$ for $n \neq m$ and hence S is not totally bounded. One can verify that the collection S is not equicontinuous at 0.

(iii) *Uniform boundedness of S:* let $X = \{0, 1\}$ and for $n \in \mathbb{N}$ define $f_n : X \to \mathbb{R}$ as $f_n(0) = 0$ and $f_n(1) = n$. Then the collection $S = \{f_n : n \in \mathbb{N}\}$ is equicontinuous but not uniformly bounded. Moreover, $D(f_n, f_m) \geq 1$ for $n \neq m$ and hence S is not totally bounded in $(C(X), D)$.

In practice, the next corollary is often referred to as the Ascoli–Arzelà Theorem. The proof immediately follows from Theorem 6.3.4, Corollary 6.2.1 and Theorem 4.2.6.

Corollary 6.3.2. *Let S be a non-empty subset of $C(X)$, where (X, d) is compact. Then S is compact in $(C(X), D)$ if and only if S is equicontinuous, closed and pointwise bounded in $(C(X), D)$.*

Exercises

6.3.1. Let (f_n) be a uniformly bounded sequence in $(C[a, b], D)$. For each n, define $\varphi_n : [a, b] \to \mathbb{R}$ by $\varphi_n(x) = \int_a^x f_n(t)dt$ for each x in $[a, b]$. Show that (φ_n) has a uniformly convergent subsequence on $[a, b]$, that is, (φ_n) has a subsequence (φ_{n_k}) such that $\varphi_{n_k} \to \varphi$ uniformly on $[a, b]$ for some $\varphi \in C[a, b]$.

6.3.2. Consider a continuous function $f : [0, \infty) \to \mathbb{R}$. For each $n \in \mathbb{N}$, define a continuous function $f_n : [0, \infty) \to \mathbb{R}$ by $f_n(x) = f(x^n)$. Show that the sequence (f_n) of continuous functions is equicontinuous at $x = 1$ if and only if f is a constant function.

6.3.3. (a) Let (X, d) be a compact metric space and let A be an equicontinuous subset of $C(X)$. Show that A is uniformly equicontinuous, that is, show that for each $\epsilon > 0$, there exists a $\delta > 0$ such that whenever x, $y \in X$ with $d(x, y) < \delta$ then $|f(x) - f(y)| < \epsilon \ \forall \ f \in A$.

(b) Let $A \subseteq C[a, b]$ such that every $f \in A$ is differentiable on (a, b). If there exists $M > 0$ such that $|f'(x)| \leq M \ \forall \ f \in A$ and $\forall \ x \in (a, b)$, then show that A is uniformly equicontinuous.

(c) Find a uniformly bounded subset F of $C(\mathbb{R})$ that is equicontinuous on $\mathbb{R}$, but not uniformly equicontinuous on $\mathbb{R}$.

(d) Give an example of any metric space (X, d) and a subset S of $C_b(X)$ such that S is uniformly bounded and uniformly equicontinuous but not totally bounded in $(C_b(X), D)$. Compare this with Theorem 6.3.4.

6.3.4. Let $X = [0, 1] \times [0, 1]$ and equip X with the Euclidean metric. Given a continuous function $f : X \to \mathbb{R}$, for each $y \in [0, 1]$ define $f_y : [0, 1] \to \mathbb{R}$ by $f_y(x) = f(x, y)$. Show that the set $F = \{f_y : y \in [0, 1]\}$ is equicontinuous on $[0, 1]$.

6.3.5. Let (X, d) be a metric space which is not necessarily compact. If (f_n) is a sequence in $C(X)$ which is equicontinuous on X and for some function $f : X \to \mathbb{R}$, we have $\lim_{n \to \infty} f_n(x) = f(x)$ for each $x \in X$, then show that f is continuous on X. In other words, the pointwise limit of an equicontinuous sequence of functions in $C(X)$ is also continuous. Compare with Theorem 6.1.2.

6.3.6. Let (X, d) be a metric space and let A be an equicontinuous subset of $C(X)$. Show that the collection F, where $F = \{x \in X : A \text{ is pointwise bounded at } x\}$, is both closed and open in (X, d).

6.3.7. Let $K : [a, b] \times [a, b] \to \mathbb{R}$ be a continuous function. Define $T : C[a, b] \to C[a, b]$ by $T(f)(x) = \int_a^b K(x, y) f(y) dy$ for each $f \in$

$C[a,b]$. If $B = \{f \in C[a,b] : \|f\|_\infty < 1\}$, then show that $T(B)$ is an equicontinuous subset of $C[a,b]$.

6.3.8. Let (X,d) be a metric space and $(f_n)_{n \in \mathbb{N}}$ be a Cauchy sequence in $(C_b(X), D)$, where D is the supremum metric. Then prove that the family $\{f_n : n \in \mathbb{N}\}$ is equicontinuous on X.

6.3.9. Prove that every uniformly convergent sequence (f_n) of real-valued bounded functions on (X,d) is uniformly bounded.

6.3.10. Let (X,d) be a compact metric space and let S be a non-empty subset of $C(X)$.

(a) If S is pointwise bounded then prove that its closure is also pointwise bounded in $C(X)$.
(b) If S is uniformly bounded then prove that its closure is also uniformly bounded in $C(X)$.
(c) Do we have similar result for equicontinuity as well?

6.4 Stone–Weierstrass Theorem

In this section, we find some nice dense subsets in $(C(X), \|\cdot\|_\infty)$, where X is a compact metric space. In 1885, the German analyst Karl Weierstrass proved that any continuous real-valued function on a closed bounded interval of $\mathbb{R}$ is the uniform limit of some sequence of polynomial functions with real coefficients. That is, given an interval $[a,b]$ and a continuous function $f : [a,b] \to \mathbb{R}$, there exists a sequence (P_n) of polynomial functions on $[a,b]$ such that (P_n) converges uniformly to f on $[a,b]$, that is, $\|P_n - f\|_\infty \to 0$ as $n \to \infty$. Since then, many different proofs of Weierstrass approximation theorem have been given. In this section, we present several versions of a vast generalization of this theorem that is due to the 20th century American analyst M. H. Stone. Before proving these versions, we would like to mention that the strategy of the proof is similar to the one employed in Ref. [55, page 146]. Let us start with some preliminary discussion which is required to attain our goal.

A linear space A of real-valued functions on a non-empty set X is called an **algebra of functions** whenever the product of any two functions in A is again in A. Thus, a set $A \subseteq \mathbb{R}^X$ is an algebra if for every pair f, $g \in A$ and real numbers α and β, we have $\alpha f + \beta g$ and fg in A, where of course, $(fg)(x) = f(x)g(x)$ for each $x \in X$.

A collection L of real-valued functions defined on a non-empty set X is said to **separate the points** of X if for every pair of distinct points x and y of X, there exists a function $f \in L$ such that $f(x) \neq f(y)$. We say that L **vanishes nowhere** on X if for each $x \in X$, there exists an $f \in L$ such that $f(x) \neq 0$.

If f and g are real-valued functions on X, we define $f \vee g$ and $f \wedge g$ on X by

$$(f \vee g)(x) = \max\{f(x), g(x)\} \quad \text{and}$$
$$(f \wedge g)(x) = \min\{f(x), g(x)\}.$$

It is easy to check that

$$f \vee g = \frac{1}{2}(f + g + |f - g|) \quad \text{and}$$

$$f \wedge g = \frac{1}{2}(f + g - |f - g|).$$

If (X, d) is a metric space and f and g are continuous on (X, d), so are $f \vee g$ and $f \wedge g$.

Now $L \subseteq \mathbb{R}^X$ is called a **lattice** (of functions) if for any two f, $g \in L$, $f \vee g$ and $f \wedge g$ belongs to L. Additionally, if L is a linear space over $\mathbb{R}$, then L is called a **function space**.

Finally, note that given a metric space (X, d), $C(X)$ is itself an algebra. Now if $\emptyset \neq L \subseteq C(X)$ and L is itself an algebra with respect to same operations as that of $C(X)$, then L is called a **subalgebra** of $C(X)$.

Proposition 6.4.1. *Let X be a non-empty set and L be an algebra of real-valued functions on X which separates the points of X and vanishes nowhere on X. If x, $y \in X$ such that $x \neq y$ and a and b are real numbers, then there is a function $h \in L$ such that $h(x) = a$ and $h(y) = b$.*

Proof. Let $x \neq y$, a and b be given. Since $x \neq y$ and L separates the points of X, there exists $f_1 \in L$ such that $f_1(x) \neq f_1(y)$. Also, since L vanishes nowhere on X, there exist f_2, $f_3 \in L$ such that $f_2(x) \neq 0 \neq f_3(y)$. Now consider the different possibilities.

Case I: Suppose $f_1(x) = 0$. Then $f_1(y) \neq 0$.

Take $h = \alpha f_1 + \beta f_2$, where $\alpha = \frac{b f_2(x) - a f_2(y)}{f_1(y) f_2(x)}$ and $\beta = \frac{a}{f_2(x)}$. Since f_1, $f_2 \in L$ and L is an algebra, $h \in L$. Moreover, $h(x) = a$ and $h(y) = b$.

Similarly, if $f_1(y) = 0$, then $f_1(x) \neq 0$ and we take $h = \alpha f_1 + \beta f_3$, where $\alpha = \frac{a f_3(y) - b f_3(x)}{f_1(x) f_3(y)}$ and $\beta = \frac{b}{f_3(y)}$. As before, $h \in L$, $h(x) = a$ and $h(y) = b$.

Case II: $f_1(x) \neq 0 \neq f_1(y)$.

Take $h = \alpha f_1 + \beta f_1^2$, where α and β are the solutions of the system,

$$\alpha f_1(x) + \beta (f_1(x))^2 = a,$$
$$\alpha f_1(y) + \beta (f_1(y))^2 = b.$$

The solutions of the system exist since the determinant of the coefficient matrix is $f_1(x) f_1(y) (f_1(y) - f_1(x)) \neq 0$. $\qquad\square$

Note that the collection of all even polynomial functions is a subalgebra of $C(\mathbb{R})$ which does not separate the points of $\mathbb{R}$. Here recall that a function $f : \mathbb{R} \to \mathbb{R}$ is said to be even if $f(x) = f(-x)$ for all $x \in \mathbb{R}$.

Theorem 6.4.1. *Let (X, d) be a compact metric space and L be a subalgebra of $C(X)$. In addition, suppose that*

(i) *L separates the points of X*
(ii) *L vanishes nowhere on X and*
(iii) *L is a lattice.*

Then given a function $g \in C(X)$, a point a in X and $\epsilon > 0$, there exists a function $f \in L$ such that $f(a) = g(a)$ and $f(x) > g(x) - \epsilon \;\; \forall \, x \in X$.

Proof. For every point x ($\neq a$) in X, by Proposition 6.4.1, there exists $h_{x,a} \in L$ such that $h_{x,a}(a) = g(a)$ and $h_{x,a}(x) = g(x) > g(x) - \epsilon$.

Note that since L vanishes nowhere on X, there exists $h \in L$ such that $h(a) \neq 0$. So if $x = a$, let $h_{a,a} = \frac{g(a)}{h(a)} h$. Since L is an algebra, $h_{a,a} \in L$. Moreover, $h_{a,a}(a) = g(a) > g(a) - \epsilon$.

Now define $V_{x,a} = \{t \in X : h_{x,a}(t) > g(t) - \epsilon\}$. Note that $h_{x,a}(x) > g(x) - \epsilon$ and $h_{x,a}(a) > g(a) - \epsilon$. Hence both x and a are in $V_{x,a}$. Also, since $h_{x,a} - g$ is continuous on X, $V_{x,a}$ is open in X. Now $\{V_{x,a} : x \in X\}$ is an open cover of the compact space X. Hence there exists a finite subset $\{x_1, \ldots, x_n\}$ of X such that $X = \bigcup_{i=1}^{n} V_{x_i,a}$. Define $f = h_{x_1,a} \vee \cdots \vee h_{x_n,a}$. Since L is a lattice, $f \in L$. Moreover, $f(a) = h_{x_i,a}(a)$ for some i, with $1 \leq i \leq n$. But $h_{x_i,a}(a) = g(a)$. So $f(a) = g(a)$. Also, if $x \in X$, then there exists some k ($1 \leq k \leq n$) such that $x \in V_{x_k,a}$. Then $f(x) \geq h_{x_k,a}(x) > g(x) - \epsilon$ and consequently, the function f satisfies the required properties. $\square$

Now we can prove the 'lattice version' of Stone–Weierstrass Theorem.

Theorem 6.4.2. *Let (X, d) be a compact metric space and L be a subalgebra of $C(X)$. In addition, suppose that*

(i) *L separates the points of X*
(ii) *L vanishes nowhere on X*
(iii) *L is a lattice.*

Then L is dense in $(C(X), \| \cdot \|_\infty)$.

Proof. Let $g \in C(X)$ and $\epsilon > 0$. For each $x \in X$, by Theorem 6.4.1, there exists a continuous function $f_x \in L$ such that $f_x(x) = g(x)$ and $f_x > g - \epsilon$.

Let $V_x = \{y \in X : f_x(y) < g(y) + \epsilon\}$. Note that since $f_x(x) = g(x) < g(x) + \epsilon$, $x \in V_x$. Moreover, since $f_x - g$ is continuous on X, V_x is open in X. Hence $\{V_x : x \in X\}$ is an open cover of the compact space X. Consequently, there exists a finite subset $\{x_1, \ldots, x_n\}$ of X such that $X = \bigcup_{i=1}^{n} V_{x_i}$. Let $f = f_{x_1} \wedge \ldots \wedge f_{x_n}$. Since L is a lattice, $f \in L$. Moreover, since $f_{x_k} > g - \epsilon$ for all $1 \leq k \leq n$, $f > g - \epsilon$. On the other hand, if $x \in X$, then there exists some k ($1 \leq k \leq n$) such that $x \in V_{x_k}$ and hence $f(x) \leq f_{x_k}(x) < g(x) + \epsilon$. Therefore, $g(x) - \epsilon < f(x) < g(x) + \epsilon \; \forall \, x \in X$ and consequently, $\|f - g\|_\infty = \sup_{x \in X} |f(x) - g(x)| \leq \epsilon$. Thus, L is dense in $(C(X), \| \cdot \|_\infty)$. $\square$

In order to state and prove the classical (algebra version) Stone–Weierstrass Theorem, we need the next two results.

Lemma 6.4.1. *There exists a sequence of polynomial functions that converges uniformly to $\sqrt{x}$ on the interval $[0, 1]$.*

Proof. First by induction, we construct a sequence of polynomial functions (P_n) as follows. Let $P_1(x) = 0$ for all $x \in [0,1]$ and then inductively define $P_{n+1}(x) = P_n(x) + \frac{1}{2}(x - (P_n(x))^2)$ for $n \geq 1$.

Now we prove by induction that $0 \leq P_n(x) \leq \sqrt{x}$ for all n and for all $x \in [0,1]$. For $n = 1$, it is trivial. Assume now $0 \leq P_n(x) \leq \sqrt{x}$ for all $x \in [0,1]$ and for some n. Then, by definition $0 \leq P_{n+1}(x)$ for all $x \in X$. Moreover, $\sqrt{x} - P_{n+1}(x) = (\sqrt{x} - P_n(x))(1 - \frac{1}{2}(\sqrt{x} + P_n(x)))$. Since $P_n(x) \leq \sqrt{x} \leq 1$, $(1 - \frac{1}{2}(\sqrt{x} + P_n(x))) \geq 0$. Hence $P_{n+1}(x) \leq \sqrt{x}$ for all $x \in [0,1]$.

Note that for each n, $P_n(x) \leq P_{n+1}(x)$ for all $x \in [0,1]$. Hence for each $x \in [0,1]$, $(P_n(x))$ is an increasing sequence in $\mathbb{R}$ which is bounded above by 1. Consequently, (P_n) converges pointwise to some non-negative function ϕ on $[0,1]$. Further, $\lim_{n\to\infty} P_{n+1}(x) = \lim_{n\to\infty} P_n(x) + \frac{1}{2}(\lim_{n\to\infty}(x - (P_n(x))^2))$, that is, $\phi(x) = \phi(x) + \frac{1}{2}(x - (\phi(x))^2)$. So $(\phi(x))^2 = x$, that is, $\phi(x) = \sqrt{x}$ for all $x \in [0,1]$ (because ϕ is non-negative).

Since ϕ is a continuous function and (P_n) is increasing, by Dini's theorem, (P_n) converges uniformly to $\sqrt{x}$ on $[0,1]$. $\qquad\square$

Theorem 6.4.3. *Let (X, d) be a metric space and A be a subalgebra of $C_b(X)$. Then the closure $\overline{A}$ of A in $(C_b(X), \|\cdot\|_\infty)$ is also an algebra. Moreover, if f and g are in $\overline{A}$, then $|f|$, $f \vee g$ and $f \wedge g$ are also in $\overline{A}$.*

Proof. Let $f, g \in \overline{A}$. Then we can choose sequences (f_n) and (g_n) in A such that $\|f - f_n\|_\infty \to 0$ and $\|g - g_n\|_\infty \to 0$ in $(C_b(X), \|\cdot\|_\infty)$. It follows that for $\alpha \in \mathbb{R}$,

$$\|(\alpha f) - (\alpha f_n)\|_\infty = |\alpha|\|f - f_n\|_\infty \to 0,$$

$$\|(f + g) - (f_n + g_n)\|_\infty \leq \|f - f_n\|_\infty + \|g - g_n\|_\infty \to 0, \quad \text{and}$$

$$\|fg - f_n g_n\|_\infty = \|f(g - g_n) + g_n(f - f_n)\|_\infty$$

$$\leq \|f\|_\infty \|g - g_n\|_\infty + \|g_n\|_\infty \|f - f_n\|_\infty$$

$$\to \|f\|_\infty . 0 + \|g\|_\infty . 0 = 0.$$

(because $|\|g_n\|_\infty - \|g\|_\infty| \leq \|g_n - g\|_\infty$, and since $\|g_n - g\|_\infty \to 0$, $\|g_n\|_\infty \to \|g\|_\infty$). Hence αf, $f + g$ and fg are all in $\overline{A}$, that is, $\overline{A}$ is also an algebra.

To prove that $|f| \in \overline{A}$, we may assume that $\beta = \|f\|_\infty > 0$. Let (P_n) be a sequence of polynomial functions (determined by Lemma 6.4.1) that converges uniformly to $\sqrt{x}$ on $[0, 1]$. Since $\overline{A}$ is an algebra, the function $\varphi_n = P_n(\frac{f^2}{\beta^2})$ belongs to $\overline{A}$ for each n. Also the sequence (φ_n) converges uniformly to $\sqrt{\frac{f^2}{\beta^2}} = \frac{|f|}{\beta}$. Then $\frac{|f|}{\beta} \in \overline{A}$ and hence $|f| \in \overline{A}$. The rest follows from

$$f \vee g = \frac{1}{2}(f + g + |f - g|) \quad \text{and}$$

$$f \wedge g = \frac{1}{2}(f + g - |f - g|). \qquad \square$$

Now we state and prove a generalized version of the classical Weierstrass approximation theorem.

Theorem 6.4.4 (Stone–Weierstrass theorem). *Let (X, d) be a compact metric space and A be a subalgebra of $C(X)$ such that A separates the points of X. Then either*

(a) *$\overline{A} = C(X)$, where $\overline{A}$ is the closure of A in $(C(X), \|\cdot\|_\infty)$ or else*
(b) *there is some $p \in X$ such that $\overline{A} = \{f \in C(X) : f(p) = 0\}$.*

Proof. *Case I:* A vanishes nowhere on X.

Then $\overline{A}$ also vanishes nowhere on X. Also note that $\overline{A}$ separates the points of X. By Theorem 6.4.3, $\overline{A}$ is an algebra as well as a lattice. Therefore, $\overline{A}$ is dense in $(C(X), \|\cdot\|_\infty)$ by Theorem 6.4.2. Since $\overline{A}$ is closed, $\overline{A} = C(X)$.

Case II: There exists some $p \in X$ such that $h(p) = 0 \ \forall \ h \in A$.

Let $L = \{c + h : c \in \mathbb{R}, \ h \in A\}$. It is easy to verify that L is a subalgebra of $C(X)$ that separates the points of X and vanishes nowhere on X. It follows from Case I that if $f \in C(X)$ with $f(p) = 0$ and $\epsilon > 0$ are given, then there is some $c + h \in L$ such that $\|f - (c + h)\|_\infty < \frac{\epsilon}{2}$. Evaluating at p, we get

$$|c| = |f(p) - (c + h(p))| \leq \|f - (c + h)\|_\infty < \frac{\epsilon}{2}.$$

Therefore, the function h in A satisfies

$$\|f - h\|_\infty \le \|f - (c+h)\|_\infty + \|c\|_\infty$$
$$= \|f - (c+h)\|_\infty + |c|$$
$$< \epsilon/2 + \epsilon/2 = \epsilon.$$

This proves that $\overline{A} = \{f \in C(X) : f(p) = 0\}$. $\qquad\square$

Note that the compactness of the metric space (X, d) cannot be dropped in Theorem 6.4.4 as seen in the following example.

Example 6.4.1. Let $X = \mathbb{R}$ and let $A = \{f \in C_b(X) : \lim\limits_{n\to\infty} f(n) \in \mathbb{R}\}$. Then A is a subalgebra of $C_b(X)$. It separates points of X: let $a \ne a'$ in X. Then $a < n_o$ for some $n_o \in \mathbb{N}$. Let $B = \{a'\} \cup \{n \in \mathbb{N} : n > n_o\}$. Now define a function f_1 on X as

$$f_1(x) = \frac{d(x, B)}{d(x, a) + d(x, B)}.$$

Then $f_1(a) \ne f_1(a')$ and $f_1 \in A$. Also, note that A contains the constant function **1**. Now we claim that $\overline{A} \ne C_b(X)$: consider the function g on X:

$$g(x) = \frac{d(x, C)}{d(x, C) + d(x, D)},$$

where $C = \{2n - 1 : n \in \mathbb{N}\}$ and $D = \{2n : n \in \mathbb{N}\}$. Then $g \in C_b(X)$ but $\sup_{x \in X} |g(x) - f(x)| \ge 1/2$ for all $f \in A$: let $f \in A$ and let $\lim\limits_{n\to\infty} f(n) = a \in \mathbb{R}$. Now consider different cases $a \le 0$, $0 < a \le 1/2$ and $a > 1/2$ and analyze the value of $|g(x) - f(x)|$ at even or odd natural numbers.

Hence, A is not dense in $(C_b(X), \|\cdot\|_\infty)$.

Corollary 6.4.1. *Let (X, d) be a compact metric space and A be a subset of $C(X)$ such that*

(a) *A is a subalgebra of $C(X)$.*
(b) *A separates the points of X.*
(c) *A contains the constant function **1**, where $\mathbf{1}(x) = 1 \ \forall \ x \in X$.*

Then A is dense in $(C(X), \|\cdot\|_\infty)$.

Now we finally state the result which is one of the finest contributions to real analysis. It follows immediately by applying Corollary 6.4.1 to the algebra of all polynomial functions (with real coefficients) defined on a compact subset of $\mathbb{R}$.

Corollary 6.4.2 (Weierstrass approximation theorem). *Any real-valued continuous function on a compact subset A of $\mathbb{R}$ is the uniform limit on A of a sequence of polynomial functions.*

Remarks. (a) Since the collection of all polynomials with rational coefficients is countable, by previous corollary the normed linear space $(C[a, b], \|\cdot\|_\infty)$ is separable. Also see Exercise 6.4.8.

(b) We do have an analogous version of this result for complex valued functions as well which says that for every complex-valued continuous function f on a compact subset A of $\mathbb{R}$, there exists a sequence of polynomials (with complex coefficients) which converges uniformly to f on A. See Exercise 6.4.10.

Further, let us prove a version of Theorem 6.4.2 which uses the hypothesis of strong separation of the points of X. This version is known as *Kakutani–Krein Theorem*. In this version, we do not require L to be a subalgebra of $C(X)$ but a function space. The precise statement is as follows.

Theorem 6.4.5 (Kakutani–Krein theorem). *Let (X, d) be a compact metric space with at least two points. Let L be a vector subspace of $C(X)$. In addition, suppose that*

(a) *L is a lattice.*

(b) *L strongly separates the points of X, that is, for each pair of distinct points x, y in X, $\bigcup\{(f(x), f(y)) : f \in L\} = \mathbb{R}^2$.*

Then L is dense in $(C(X), \|\cdot\|_\infty)$.

Proof. Let $g \in C(X)$ and $\epsilon > 0$. We need to find $f \in L$ such that $g(x) - \epsilon \le f(x) \le g(x) + \epsilon \ \forall \ x \in X$. By the strong separation property, for any x, $y \in X$ (x and y need not be distinct), we can find $h_{x,y} \in L$ such that $h_{x,y}(x) = g(x)$ and $h_{x,y}(y) = g(y)$. (Here we are using the fact that X has at least two points.)

Fix $x_o \in X$. For each $y \in X$, define $U_y = \{z \in X : h_{x_o,y}(z) < g(z) + \epsilon\}$. Since $h_{x_o,y}(y) = g(y) < g(y) + \epsilon$, $y \in U_y$. Moreover, since

$h_{x_o,y} - g$ is continuous on X, U_y is open in X. Hence $\{U_y : y \in X\}$ is an open cover of the compact space X. Consequently, there exists a finite subset $\{y_1, \ldots, y_n\}$ of X such that $X = \bigcup_{i=1}^{n} U_{y_i}$.

Let $f_{x_o} = h_{x_o,y_1} \wedge \ldots \wedge h_{x_o,y_n}$. Since L is a lattice, $f_{x_o} \in L$. Note that $f_{x_o}(x_o) = g(x_o)$. If $z \in X$, then $z \in U_{y_j}$ for some j, $1 \le j \le n$. Hence $f_{x_o}(z) \le h_{x_o,y_j}(z) < g(z) + \epsilon$. Since $x_o \in X$ was arbitrary, we have $f_x(z) < g(z) + \epsilon \ \forall \ z \in X$ and $\forall \ x \in X$ $(\star)$.

Now we repeat this argument 'upside down'. Given $x \in X$, define $V_x = \{z \in X : f_x(z) > g(z) - \epsilon\}$. Since $f_x(x) = g(x) > g(x) - \epsilon$, $x \in V_x$. Again, since $f_x - g$ is continuous on X, V_x is open in X. Hence $\{V_x : x \in X\}$ is an open cover of the compact space X. Therefore, there exists a finite subset $\{x_1, \ldots, x_m\}$ of X such that $X = \bigcup_{i=1}^{m} V_{x_i}$.

Let $f = f_{x_1} \vee \ldots \vee f_{x_m}$. Since L is a lattice, $f \in L$. If $z \in X$, then $z \in V_{x_k}$ for some k, $1 \le k \le m$. Hence $f(z) \ge f_{x_k}(z) > g(z) - \epsilon$. Thus, $f(z) > g(z) - \epsilon \ \forall \ z \in X$ $(\star\star)$.

By $(\star)$, we have $f_{x_j}(z) < g(z) + \epsilon$ for each $z \in X$ and for each $j \in \{1, \ldots, m\}$. Therefore, $f(z) < g(z) + \epsilon \ \forall \ z \in X$ $(\star\star\star)$.

By $(\star\star)$ and $(\star\star\star)$, we get $g(z) - \epsilon < f(z) < g(z) + \epsilon \ \forall \ z \in X$. Thus, we have constructed the required f in L. $\qquad\square$

Remark. If X contains only one point, then $L = \{0\}$ satisfies the hypothesis of previous theorem but L is not dense in $(C(X), \|\cdot\|_\infty)$.

Evidently, if a subcollection A of $\mathbb{R}^X$ strongly separates the points of X, then it separates the points of X. But the converse is not true in general. For example, if $X = \{a, b\}$ is a set of two distinct points and $A = \{f\}$ where $f(a) = 1$ and $f(b) = 0$. Then clearly A separates the points of X but not strongly. Also see Exercise 6.4.2.

Historical Perspective

The idea of pointwise convergence of a sequence of real valued functions existed since the early days of calculus, particularly in the study of power series and trigonometric series. But the uniform convergence of sequences of functions could not have been even imagined, until the concepts of convergent series and continuous function had been precisely described by Bolzano and Cauchy. It suffices to mention that before them, the use of series without regard to convergence and divergence had led to a number of paradoxes and disagreements.

In 1821, while studying the limit of a convergent series of continuous functions and the term by term integration of a series of continuous functions, Cauchy made some missteps and overlooked the need for uniform convergence. Fortunately soon after, Cauchy's errors came to the notice of Abel. In his 1826 paper, Abel gave a correct proof for sum of a uniformly convergent series of continuous functions to be continuous in the interior of the interval of convergence. But he did not study the uniform convergence of a series of functions in its generality. The notion of uniform convergence and its subsequent importance were recognized in and of themselves by Stokes, a leading mathematical physicist of his time and independently by Philipp L. Seidel in 1847–1848 and by Cauchy himself in 1853.

Actually Weierstrass had the precise idea of uniform convergence with perfect clarity as early as 1842. But his work related to uniform convergence was first published much later in 1894. In fact, according to Stephen Willard ([58], p. 320), 'In the last half of the 19th century, in the hands of Heine, Weierstrass, Riemann and others, uniform convergence came into its own in applications to integration theory and Fourier series'. Apparently, the work of Ascoli, Arzelà and Hadamard in the last two decades of the 19th century marked the beginning of what is known today as theory of function spaces. Loosely speaking, a topological space in which the points are functions is called a function space. The spaces of functions have been used since the late 19th century to form a framework in which convergence of sequences of functions could be studied.

During his years as a high school teacher, Weierstrass discovered that any real-valued continuous function over a closed bounded interval in the real line can be expressed in that interval as the uniformly convergent limit of a sequence of polynomials. This result, better known as uniform approximation of real-valued continuous functions on a closed bounded interval, proved to be a strong and useful tool in classical analysis. In the middle of 20th century, M. H. Stone enriched the theory of approximation of continuous functions by generalizing the aforesaid approximation theorem of Weierstrass substantially to real or complex valued functions having any compact Hausdorff space as domain.

Note. *For the historical perspective, the authors have taken substantial help from Ref. [38].*

Exercises

6.4.1. This exercise gives a 'lattice' version of Stone–Weierstrass Theorem under a different set of hypothesis. Prove the following results:

(a) Let X be a non-empty set and L be a vector space of real-valued functions on X that separates the points of X and contains the constant function 1. Then given any two distinct points x and y of X and real numbers α and β, there exists some $f \in L$ such that $f(x) = \alpha$ and $f(y) = \beta$.
(b) Let (X, d) be a compact metric space and let L be a function space of continuous functions on X that contains the constant function 1 and separates the points of X. Then

  (i) Given $g \in C(X)$, a point a in X and $\epsilon > 0$, there exists a function $f \in L$ such that $f(a) = g(a)$ and $f(x) > g(x) - \epsilon \ \forall \, x \in X$.
  (ii) Moreover, L is dense in $(C(X), \|\cdot\|_\infty)$.

6.4.2. Let S be a linear subspace of $C(X)$, where (X, d) is a metric space. Suppose that S contains the constant function 1 and S separates the points of X. Show that S strongly separates the points of X.

6.4.3. Let (X, d) be a compact metric space and S be a subalgebra of $C(X)$ such that S is closed in $(C(X), \|\cdot\|_\infty)$. Show that S is a lattice.

6.4.4. Suppose $\mathrm{Lip}(X)$ denotes the collection of all real-valued Lipschitz functions on a metric space (X, d). Prove that

(a) $\mathrm{Lip}(X)$ is a function space.
(b) $\mathrm{Lip}(X)$ is a subalgebra of $C(X)$ provided (X, d) is compact.

6.4.5. Show that $\mathrm{Lip}(X)$ is dense in $(C(X), \|\cdot\|_\infty)$ provided (X, d) is compact.

6.4.6. Let (X, d) be a compact metric space and L be a lattice of continuous functions on X. Suppose $f \in C(X)$. Given x, $y \in X$ and $\epsilon > 0$, there exists $g \in L$ such that $|f(x) - g(x)| < \epsilon$ and $|f(y) - g(y)| < \epsilon$. Show that $f \in \overline{L}$, where $\overline{L}$ is the closure of L in $(C(X), \|\cdot\|_\infty)$.

6.4.7. If f is a real-valued continuous function on $[0, 1]$ such that $\int_0^1 x^n f(x)dx = 0$ for $n = 0, 1, \ldots$, then show that $f(x) = 0 \ \forall \ x \in [0, 1]$.

6.4.8. If (X, d) is a compact metric space, then show that $(C(X), D)$ is separable, where D is the supremum metric.

6.4.9. (a) Give an example of a sequence of differentiable functions, defined on some bounded interval of $\mathbb{R}$, which is uniformly convergent to a non-differentiable function. Compare with Theorem 6.1.2.

(b) With the help of an example, show that if (f_n) is a sequence of differentiable functions which converges uniformly to a differentiable function f, then this does not imply that the sequence of derivatives (f_n') converges pointwise to f'.

(c) For all $n \in \mathbb{N}$, let $f_n : I \to \mathbb{R}$ be differentiable, where I is some bounded interval of $\mathbb{R}$, such that the real sequence $(f_n(x_o))$ converges for some $x_o \in I$. If $f_n' \to g$ uniformly, then prove that the sequence (f_n) converges uniformly to a function $f : I \to \mathbb{R}$ that is differentiable on I and $f' = g$.

6.4.10 (Stone–Weierstrass theorem for complex valued functions). Let (X, d) be a compact metric space and A be a self-adjoint subalgebra of $C(X, \mathbb{C})$, where $C(X, \mathbb{C})$ denotes the set of all continuous functions from (X, d) to $\mathbb{C}$ ($\mathbb{C}$ is equipped with the usual distance metric). If A separates the points of X, then prove that either

(a) $\overline{A} = C(X, \mathbb{C})$, where $\overline{A}$ is the closure of A in $(C(X, \mathbb{C}), \|\cdot\|_\infty)$ or else

(b) there is some $p \in X$ such that $\overline{A} = \{f \in C(X, \mathbb{C}) : f(p) = 0\}$.

Here note that A is said to be self-adjoint if for every $f \in A$, its complex conjugate $\overline{f}$ also lies in A, where $\overline{f}(x) = \overline{f(x)}$.

Chapter 7

Connectedness

Like completeness and compactness, connectedness is another crucial pillar in the analysis of metric spaces. The well-known Intermediate Value Theorem, that is highly used in calculus and numerical analysis, is not only based on continuity of a function but also it makes use of the connectedness of an interval in $\mathbb{R}$. In fact, many of the results in real analysis are given for intervals because implicitly we use their connectedness feature as well. Thus, our study of metric spaces would be incomplete without the discussion of connected metric spaces.

7.1 Connected Metric Spaces

Intuitively, we say that something is connected if it consists of one single piece. In this section, we plan to make this notion more precise. An interval in $\mathbb{R}$ has the property that if it contains two distinct points, then it contains every point between them. Connectedness represents an extension of this idea that an interval consists of one piece. The problem of deciding whether a metric space is in one piece is resolved by determining whether it can be split into non-empty disjoint open subsets.

Definitions 7.1.1. A metric space (X, d) is called *connected* if it cannot be expressed as a union of two non-empty disjoint open subsets of (X, d).

If a metric space (X, d) is not connected then it is referred to as **disconnected** metric space.

Remark. By the definition of connected metric space, it is clear that the metric structure of (X, d) is not used in full strength. In fact, it makes use of open sets in (X, d). Consequently, the notion of connectedness can be defined for any topological space.

Note that proving some metric space to be disconnected is easier than proving it to be connected. This is because for disconnectedness, you just need to produce two non-empty disjoint open subsets A and B of (X, d) such that $X = A \cup B$. But for connectedness, you need to prove that no such open subsets exist. Now that is a task! Consequently, there is a need to learn some equivalent characterizations of connectedness which are easier to apply.

Definition 7.1.2. A non-empty subset Y of a metric space (X, d) is called *connected* if the metric space (Y, d) is itself connected. In this case, we say that Y is connected in (X, d).

Because of the definition of connected subsets in a metric space (X, d), one needs to be conscious of the ambient space while considering open and closed sets.

Examples 7.1.1. (a) Intervals in $\mathbb{R}$ are connected (Theorem 7.1.2) and hence $\mathbb{R} = (-\infty, \infty)$ is also connected. But $\mathbb{R} \setminus \{0\}$ is disconnected because $\mathbb{R} \setminus \{0\} = (-\infty, 0) \cup (0, \infty)$.
(b) Every metric space (with at least two points) in which every point is an isolated point is disconnected.
(c) The subset $A = \{(x, y) \in \mathbb{R}^2 : xy \neq 0\}$ of $\mathbb{R}^2$ is disconnected because $xy \neq 0$ means both x and y are non-zero. Note that $A = \mathbb{R}^2 \setminus B$ where $B = \{(x, 0), (0, y) : x, \ y \in \mathbb{R}\}$. *Now geometrically can you see the possible non-empty disjoint open subsets of A such that A could be expressed as their unions?*

Now let us discuss some nice characterizations of a connected metric space which can be applied according to the convenience.

Theorem 7.1.1. *Let (X, d) be a metric space. Then the following assertions are equivalent:*

(a) *(X, d) is connected.*
(b) *X is not a union of two disjoint non-empty closed subsets.*
(c) *Every continuous function from (X, d) to $\{0, 1\}$ is constant, where $\{0, 1\}$ is equipped with the discrete metric.*

(d) *Every proper non-empty subset of X has non-empty boundary in* (X, d).

(e) *No proper non-empty subset of X is both open and closed in* (X, d).

Proof. (a) $\Rightarrow$ (b): This is immediate.

(b) $\Rightarrow$ (c): If possible, suppose there is a non-constant continuous function $f : (X, d) \to \{0, 1\}$. Then $X = f^{-1}\{0\} \cup f^{-1}\{1\}$. Since f is non-constant, both $f^{-1}\{0\}$ and $f^{-1}\{1\}$ are non-empty. Since f is continuous and $\{0\}$, $\{1\}$ are closed in the range space, both $f^{-1}\{0\}$ and $f^{-1}\{1\}$ are closed in (X, d) by Theorem 2.1.1. But this contradicts (b). Hence our supposition was wrong.

(c) $\Rightarrow$ (d): Let A be a non-empty proper subset of X. If possible, suppose that the boundary of A in X, $\partial(A)$ is empty. Thus, $\partial(A) = \overline{A} \cap \overline{X - A} = \emptyset$. Consequently, $\overline{A} \subseteq A$ and $\overline{X - A} \subseteq X - A$ which implies that A and $X - A$ are closed in X. Define $f : X \to \{0, 1\}$ as: $f(x) = 1$ if $x \in A$, and $f(x) = 0$ if $x \in X - A$. Thus $f = \chi_A$, the characteristic function of A. Note that the closed sets in $\{0, 1\}$ are $\emptyset, \{0\}, \{1\}$ and $\{0, 1\}$. By construction, the inverse image of each of these sets is closed in (X, d). Hence f is continuous by Theorem 2.1.1. But f is non-constant and this contradicts (c). Hence $\partial(A) \neq \emptyset$.

(d) $\Rightarrow$ (e): If possible, suppose A is a proper non-empty subset of X which is both open and closed in (X, d). Then $X - A$ is also closed in (X, d) and consequently, $\partial A = \overline{A} \cap \overline{(X - A)} = A \cap (X - A) = \emptyset$. But this contradicts (d).

(e) $\Rightarrow$ (a): This is immediate. $\qquad\qquad\qquad\qquad\qquad\qquad\square$

It is now evident that the set of rational numbers $\mathbb{Q}$ is not connected because $(-\pi, \pi) \cap \mathbb{Q}$ is a clopen subset of $\mathbb{Q}$. It is known that the continuous image of a compact metric space is compact. The next result talks about the analogous result for connected spaces.

Corollary 7.1.1. *Let* $f : (X, d) \to (Y, \rho)$ *be a continuous map between two metric spaces. If A is a non-empty connected subset of* (X, d), *then $f(A)$ is connected in* (Y, ρ).

Proof. If $(f(A), \rho)$ is not connected, then by Theorem 7.1.1, there exists a non-constant continuous function $g : (f(A), \rho) \to \{0, 1\}$. But then the composite $g \circ f : (A, d) \to \{0, 1\}$ will be a non-constant continuous function, which in turn will imply that (A, d) is not connected. A contradiction. Hence $f(A)$ is connected in (Y, ρ). □

What can you say about the converse of the previous corollary? (see Exercise 7.1.7.)

Example 7.1.2. Let $S^1 = \{(x, y) \in \mathbb{R}^2 : x^2 + y^2 = 1\}$ be the unit circle in $\mathbb{R}^2$. By Theorem 2.2.4 (and the remark afterwards), the function $f : [0, 2\pi) \to S^1$ defined as $f(x) = (\cos x, \sin x)$ is continuous. By Corollary 7.1.1, $S^1 = f([0, 2\pi))$ is connected.

Corollary 7.1.1 says that connectedness is a topological property, that is, it is preserved under homeomorphism. Although it cannot be used in proving two spaces to be homeomorphic but at least it may help in disproving two spaces to be homeomorphic.

Example 7.1.3. The interval $[a, b)$ is not homeomorphic to (a, b): suppose, if possible, $f : [a, b) \to (a, b)$ is a homeomorphism. Then note that (a, b) is a connected subset of $[a, b)$, while $f((a, b)) = (a, b) \setminus \{f(a)\}$ is not connected because $(a, f(a))$ is a non-empty proper subset of $(a, b) \setminus \{f(a)\}$ which is both open and closed in $(a, b) \setminus \{f(a)\}$. This gives a contradiction.

Corollary 7.1.2. *Let (X, d) be a metric space and A be a connected subset of X. Suppose $A \subseteq B \subseteq \overline{A}$. Then B is also connected in (X, d).*

Proof. Let $f : (B, d) \to \{0, 1\}$ be a continuous function. Then $f|_A$ is also continuous. Since A is connected in (X, d), $f|_A$ is constant. Now let $x \in B \subseteq \overline{A}$, then there exists a sequence (x_n) in A such that $x_n \to x$ in (X, d). By the continuity of f, $f(x_n) \to f(x)$ in $\{0, 1\}$. But each $x_n \in A$ and $f|_A$ is constant. Hence $(f(x_n))$ is a constant sequence. So f has the same value at x as it has on A. Since x was chosen arbitrarily from B, f is also constant on B and hence (B, d) is connected, that is, B is connected in (X, d). □

As a consequence of the previous result, it is clear that the closure of a connected set in a metric space is also connected. Now the next corollary highlights how connectedness fills the gap between UCness and compactness.

Corollary 7.1.3. *Let* (X, d) *be a connected metric space. Then* (X, d) *is compact if it is a UC space.*

Proof. First we claim that $X' = X$. Suppose, if possible, $X' \neq X$. Then there exists $x \in X \setminus X'$. Hence we can find an $\epsilon > 0$ such that $B(x, \epsilon) = \{x\}$. Consequently, $\{x\}$ is open in X. Moreover, we know that finite sets are closed in a metric space. Hence $\{x\}$ is also closed in X. Then by Theorem 7.1.1(e), (X, d) is not connected (note that if $X = \{x\}$ then X is obviously compact). This gives a contradiction. Hence $X' = X$. Since (X, d) is a UC space, by Corollary 5.2.6 we get the compactness of (X, d). $\square$

Remark. Consequently, one should look for a non-connected metric space in order to find examples of UC spaces which are not compact.

In the beginning of this chapter, it was mentioned that intervals in $\mathbb{R}$ are connected. Now let us prove it precisely. Surprisingly, these are the only connected subsets of $\mathbb{R}$.

Theorem 7.1.2. *A non-empty subset I of $\mathbb{R}$ with at least two points is connected in* $(\mathbb{R}, |\cdot|)$ *if and only if I is an interval.*

Proof. First, let I be a non-empty connected subset of $\mathbb{R}$. If possible, assume that I is not an interval. So there exist a, b in I and $x \in \mathbb{R}$ such that $a < x < b$ but $x \notin I$. Then $I \cap (-\infty, x)$ and $I \cap (x, \infty)$ are both relatively open in I. Moreover, $a \in I \cap (-\infty, x)$ and $b \in I \cap (x, \infty)$. But $I = (I \cap (-\infty, x)) \cup (I \cap (x, \infty))$ which implies that I is not connected in $(\mathbb{R}, |\cdot|)$. A contradiction. Hence I must be an interval.

Conversely, suppose that I is an interval in $\mathbb{R}$ having more than one point. If I is not connected in $\mathbb{R}$, then by Theorem 7.1.1, there exists a continuous surjection $f : I \to \{0, 1\}$. Pick $a, b \in I$ such that $f(a) = 0$ and $f(b) = 1$. We may assume that $a < b$ and hence $[a, b] \subseteq I$ as I is an interval. Let $B = \{c \in [a, b) : f(c) = 0\}$. Note that $cl_I(B) \subseteq [a, b] \subseteq I$. Now $a \in B \Rightarrow B$ is non-empty and $B \subseteq [a, b] \Rightarrow B$ is bounded. Hence $c_o = \sup B \in \mathbb{R}$ and $c_o \leq b$. Since $c_o \in \overline{B}$ and $f(c) = 0 \ \forall \ c \in B$, by the continuity of f, $f(c_o) = 0$. Then $c_o < b$ because $f(b) = 1$. Also $f(x) = 1 \ \forall \ x \in (c_o, b]$. Since $c_o \in [c_o, b] = \overline{(c_o, b]}$ (the closure of $(c_o, b]$ in I), $f(c_o) = 1$ by the continuity of f. We arrive at a contradiction and consequently, I is connected in $(\mathbb{R}, |\cdot|)$. $\square$

Remark. In the converse part of Theorem 7.1.2, it was assumed that $a < b$. *Can you prove the result for $a > b$?* (Exercise 7.1.8.)

As a consequence of the previous result, $\mathbb{R} = (-\infty, \infty)$ is connected. Hence $\mathbb{R}$ has no non-trivial clopen subsets.

Corollary 7.1.4 (Intermediate value theorem). *Let f be a real-valued continuous function on a connected metric space (X, d). If a, b are in X such that $f(a) < f(b)$, then for each d in $\mathbb{R}$ with $f(a) < d < f(b)$, there exists some $c \in X$ such that $f(c) = d$.*

Proof. By Corollary 7.1.1, $f(X)$ is connected in $\mathbb{R}$. Then Theorem 7.1.2 implies that $f(X)$ is an interval in $\mathbb{R}$ and hence $[f(a), f(b)] \subseteq f(X)$.   □

The next result is a special case of the previous corollary. It is widely applicable, especially in numerical analysis for getting approximate location of roots of a function. Hence, we state the result in a precise manner.

Corollary 7.1.5 (Intermediate value theorem of calculus). *If f is a real-valued continuous function on an interval I, then f has the Intermediate Value Property (IVP) : whenever a, $b \in I$, $a < b$ and y lies strictly between $f(a)$ and $f(b)$, then there exists $x \in (a, b)$ such that $f(x) = y$.*

Remarks. (a) Note that IVP essentially says that the image of an interval under a real-valued continuous function is again an interval.

(b) In the previous corollary, if we particularly take y to be 0, then according to the result there exists $x \in (a, b)$ such that $f(x) = 0$.

Now as a consequence of this intermediate value theorem, we have the following nice observation (compare with Exercise 4.3.5). This result finds applications in various branches of mathematics including numerical analysis and fixed point theory.

Corollary 7.1.6. *Let $f : ([a, b], |\cdot|) \to ([a, b], |\cdot|)$ be a continuous function. Then f has a fixed point, that is, there is an $x_o \in [a, b]$ such that $f(x_o) = x_o$.*

Proof. Consider the function $g : [a, b] \to \mathbb{R}$ defined as $g(x) = f(x) - x$. Then g is continuous on $[a, b]$. Since $a \le f(x) \le b$ for all $x \in [a, b]$,

$g(a) \geq 0$ and $g(b) \leq 0$. Thus, $g(b) \leq 0 \leq g(a)$. By intermediate value theorem, there exists $x_o \in [a, b]$ such that $g(x_o) = 0$. This implies that $f(x_o) = x_o$. □

Geometrically, observe that a fixed point of a real-valued function represents the point of intersection of the graph of the function with the line $y = x$. And to get the intuition for the previous result, one should try to draw a continuous function $f : [0, 1] \to [0, 1]$ without crossing the line $y = x$!

Exercises

7.1.1. Give an example of a continuous function $f : [0, 1) \to [0, 1)$ which does not have any fixed point.

7.1.2. Give an example to show that the interior of a connected set need not be connected.

7.1.3. Show that if $a > 0$ and $m \in \mathbb{N}$, then a has a positive mth root, that is, there exists b in $(0, \infty)$ such that $b^m = a$.

7.1.4. Show that a connected subset of a metric space is either singleton or uncountable. Hence a finite subset of a metric space is connected if and only if it contains precisely one point.

7.1.5. Show that if every real-valued continuous function f on (X, d) has intermediate value property (that is, if a, $b \in X$ and $y \in \mathbb{R}$ lies in between $f(a)$ and $f(b)$, then there exists $x \in X$ such that $f(x) = y$), then (X, d) is connected.

7.1.6. Prove that an interval I in $\mathbb{R}$ is not homeomorphic to the unit circle $S^1 = \{(x, y) \in \mathbb{R}^2 : x^2 + y^2 = 1\}$ in $\mathbb{R}^2$. Further, if $f : S^1 \to I$ is continuous and one-to-one then show that f cannot be onto.

7.1.7. Suppose a function $f : (X, d) \to (Y, \rho)$ preserves connectedness, that is, if A is a non-empty connected subset of (X, d) then $f(A)$ is connected in (Y, ρ). Is the function f continuous?

7.1.8. Let I be a subset of $\mathbb{R}$. Suppose $f : I \to \{0, 1\}$ is a continuous function and $a, b \in I$ with $a > b$ and $f(a) = 0$, $f(b) = 1$. Show that I cannot be an interval.

7.2   More About Connectedness

Given a metric space (X, d) and a family $\{X_i : i \in I\}$ of connected subsets of X, now we study the conditions under which $\bigcup_{i \in I} X_i$ is connected. But we would like to study in a broader perspective of chained collection of sets. So first let us define this collection.

Definition 7.2.1. Let X be a non-empty set and $\{X_i : i \in I\}$ be a non-empty family of non-empty subsets of X. The family $\{X_i : i \in I\}$ is called **chained** if given X_i, X_j, i, $j \in I$, there exists $\{i_1, i_2, \ldots, i_k\} \subseteq I$ such that $X_i = X_{i_1}, X_j = X_{i_k}$ and $X_{i_{m-1}} \cap X_{i_m} \neq \emptyset$ for all $m \in \{2, 3, \ldots, k\}$. Such a k-tuple $\{X_{i_1}, X_{i_2}, \ldots, X_{i_k}\}$ is called a **chain** from X_i to X_j in $\{X_k : k \in I\}$.

Consider any two disjoint open discs in $\mathbb{R}^2$ then it is evident that the discs are connected but their union is not connected in $\mathbb{R}^2$. Here note that the family of two discs is not chained.

Theorem 7.2.1. *Let (X, d) be a metric space and $\{X_i : i \in I\}$ be a chained collection of connected subsets of X. Then $\bigcup_{i \in I} X_i$ is also connected in (X, d).*

Proof. Suppose $f : \bigcup_{i \in I} X_i \to \{0, 1\}$ is a continuous function. Since each X_i is connected in (X, d), $f|_{X_i}$ is constant for all $i \in I$ by Theorem 7.1.1. Let x, $y \in \bigcup_{i \in I} X_i$. Suppose $x \in X_i$, $y \in X_j$ for some i, $j \in I$. If $X_i = X_j$, then since $f|_{X_j}$ constant, we have $f(x) = f(y)$. If $X_i \neq X_j$, then there exists a natural number $k > 1$ and a chain $\{X_{i_1}, X_{i_2}, \ldots, X_{i_k}\}$ from X_i to X_j in $\{X_l : l \in I\}$. For each $m \in \{2, 3, \ldots, k\}$, pick $x_m \in X_{i_{m-1}} \cap X_{i_m}$. Since $f|_{X_l}$ is constant for each $l \in \{1, 2, \ldots, k\}$, we have $f(x) = f(x_2)$, $f(x_{l-1}) = f(x_l)$ for each $l \in \{3, 4, \ldots, k\}$ and $f(x_k) = f(y)$. Hence $f(x) = f(y)$ and so f is constant on $\bigcup_{i \in I} X_i$. Consequently, $\bigcup_{i \in I} X_i$ is connected in (X, d) by Theorem 7.1.1. $\qquad\square$

Now let us apply the previous result on some particular chained collections.

Corollary 7.2.1. *Let $\{X_i : i \in I\}$ be a non-empty family of connected subsets of a metric space (X, d). If $\bigcap_{i \in I} X_i \neq \emptyset$, then $\bigcup_{i \in I} X_i$ is connected in (X, d).*

Proof. The collection $\{X_i : i \in I\}$ is chained because for any $i,\ j \in I$, $X_i \cap X_j \neq \emptyset$. Hence $\bigcup_{i \in I} X_i$ is connected in (X, d) by Theorem 7.2.1. $\qquad\square$

Corollary 7.2.2. *Let (X, d) be a metric space and suppose for each pair of points x, y in X, there exists a connected subset E_{xy} of X which contains x and y. Then (X, d) is connected.*

Proof. Fix $a \in X$. Then $X = \bigcup_{x \in X} E_{ax}$ and $a \in \bigcap_{x \in X} E_{ax}$. Hence by Corollary 7.2.1, (X, d) is connected. $\qquad\square$

Corollary 7.2.3. *Let (X, d) be a metric space such that $X = \bigcup_{n \in \mathbb{N}} X_n$, where each X_n is connected and $X_{n-1} \cap X_n \neq \emptyset$ for all $n \geq 2$. Then (X, d) is connected.*

Proof. Clearly $\{X_n : n \in \mathbb{N}\}$ is a chained collection of connected subsets of X. $\qquad\square$

Example 7.2.1. The Euclidean space $\mathbb{R}^n$ is connected: since $(\mathbb{R}, |\cdot|)$ is connected and a continuous image of a connected metric space is connected (Corollary 7.1.1), every straight line passing through the origin in $\mathbb{R}^n$ is connected (consider the function $f : \mathbb{R} \to \mathbb{R}^n$ defined as $f(t) = (a_1 t, a_2 t, \ldots, a_n t)$ where $(a_1, a_2, \ldots, a_n) \in \mathbb{R}^n$ is fixed). Hence by Corollary 7.2.1, $(\mathbb{R}^n, |\cdot|)$ is connected as $\mathbb{R}^n$ is the union of all straight lines passing through the origin.

Since we talked about the connectedness of union of connected sets, now one might be thinking about the non-empty intersection of connected sets as well. In general, it is not connected.

Example 7.2.2. Consider the unit circle in $\mathbb{R}^2$ with centre as origin and the set $\{(x, 0) \in \mathbb{R}^2 : -1 \leq x \leq 1\}$. Then both are connected in $\mathbb{R}^2$ but their intersection $\{(-1, 0), (1, 0)\}$ is not connected.

By looking at the crucial role being played by connectedness in various branches of mathematics, whenever a metric space (X, d) is itself not connected, then we should look at the maximal connected subsets of X. In order to do this, we need to define a connected component in (X, d).

Definition 7.2.2. Let (X, d) be a metric space and $\emptyset \neq U \subseteq X$. Then U is called a **connected component** (or simply a **component**) of X if the following conditions are satisfied:

(i) U is connected in (X, d).
(ii) Whenever $U \subseteq V \subseteq X$ such that V is connected in (X, d), then $V = U$.

In other words, U is called a connected component of X if U is a maximal connected subset of X. Here of course, maximality is with respect to set inclusion.

Examples 7.2.3. (a) If (X, d) is itself connected then X is the only connected component of X.
(b) In a metric space (X, d) in which every point is isolated, the connected components are precisely its singleton subsets because each singleton subset of X is both open and closed in (X, d).

Let us observe a few interesting facts regarding connected components of a metric space.

Theorem 7.2.2. *Let (X, d) be a metric space. Then*

(a) *the distinct connected components of X are pairwise disjoint.*
(b) *the connected components of X are closed in (X, d).*
(c) *every connected subset of X is contained in a connected component of X.*
(d) *X is the union of its connected components.*

Proof. (a) Let C and D be two distinct connected components of X. If possible, suppose $C \cap D \neq \emptyset$. Then by Corollary 7.2.1, $C \cup D$ is connected. So by the definition of a connected component, $C = C \cup D = D$. But this is a contradiction as C and D are distinct. Hence $C \cap D = \emptyset$.

(b) Let C be a connected component of X. Then $\overline{C}$ is also connected in (X, d) by Corollary 7.1.2. Then the definition of a connected component implies that $C = \overline{C}$ and consequently, C is closed in (X, d).

(c) Let D be a connected subset of X and let $\mathcal{C}_D$ be the collection of all connected subsets of X which contains D. Define $C = \cup \{A : A \in \mathcal{C}_D\}$. Note that $\mathcal{C}_D$ is non-empty as $D \in \mathcal{C}_D$ and hence the definition of C makes sense. Then by Corollary 7.2.1, C is connected in (X, d). If E is a connected subset of X containing

C, then $D \subseteq E$. But then $E \in \mathcal{C}_D$ and hence $E \subseteq C$. Therefore, $E = C$ and consequently C is a connected component of X which contains D.

(d) Note that for each $x \in X$, $\{x\}$ is connected in (X, d). Hence by (c), x belongs to a connected component, say C_x, of X. Then $X = \cup\{C_x : x \in X\}$. $\qquad\square$

Remark. A component in a metric space need not be open. Consider the metric subspace $X = \{\frac{1}{n} : n \in \mathbb{N}\} \cup \{0\}$ of $(\mathbb{R}, |\cdot|)$. The set $\{0\}$ is a connected component of X, but $\{0\}$ is not open in $(X, |\cdot|)$ because every open interval around 0 contains points of the form $\frac{1}{n}$ eventually.

Now we would like to give a useful characterization of disconnected subsets of a metric space in terms of separated sets. But before we introduce another terminology, let us first make an observation. Consider the following two subsets of $\mathbb{R}^2$: $A = \{(x, y) : x^2 + y^2 < 1\}$ and $B = \{(x, y) : 1 \le x^2 + y^2 < 2\}$. Then $A \cup B$ is connected in $\mathbb{R}^2$. But $A \cup (B \setminus \partial B)$ is disconnected. Now try to observe this closely. Here $A \cap B = \emptyset$ but $\overline{A} \cap B \ne \emptyset$. This feature gives an idea for the kind of separation needed for a subset to be disconnected.

Definition 7.2.3. Let A and B be two non-empty subsets of a metric space (X, d). Then A and B are referred to as **separated** in (X, d) if $\overline{A} \cap B = \emptyset = A \cap \overline{B}$.

For proving the required characterization of disconnected metric space, we need to prove a small result which highlights the gap between separated sets in a metric space.

Lemma 7.2.1. *Let (X, d) be a metric space and A and B be two non-empty separated sets in (X, d). Then there exist disjoint open sets U and V in (X, d) such that $A \subseteq U$ and $B \subseteq V$.*

Proof. If $\overline{A} \cap \overline{B} = \emptyset$, then the result follows from Corollary 2.2.2. So we can assume that $\overline{A} \cap \overline{B} \ne \emptyset$. Consider $Y = X - (\overline{A} \cap \overline{B})$. Then Y is open in (X, d). Since A and B are separated in X, $A \subseteq Y$ and $B \subseteq Y$. Moreover, the closure of A in Y, $\mathrm{cl}_Y A = \overline{A} \cap Y$. Similarly, $\mathrm{cl}_Y B = \overline{B} \cap Y$. Now $(\mathrm{cl}_Y A) \cap (\mathrm{cl}_Y B) = (\overline{A} \cap \overline{B}) \cap Y = \emptyset$. Hence by applying Corollary 2.2.2 to the metric space (Y, d), we get disjoint and relatively open sets U and V in (Y, d) such that $A \subseteq U$ and $B \subseteq V$. But since Y itself is open in (X, d), U and V are also open in (X, d). $\qquad\square$

Theorem 7.2.3. *Let (X, d) be a metric space and $\emptyset \neq Y \subseteq X$. Then the following statements are equivalent:*

(a) *(Y, d) is disconnected.*
(b) *There exist non-empty separated sets A and B in (X, d) such that $Y = A \cup B$.*
(c) *There exist disjoint open sets U and V in (X, d) such that $Y \subseteq U \cup V$, $U \cap Y \neq \emptyset$ and $V \cap Y \neq \emptyset$.*

Proof. (a) $\Rightarrow$ (b): Since (Y, d) is disconnected, there exist relatively open sets A and B in Y such that $A \neq \emptyset \neq B$, $A \cap B = \emptyset$ and $Y = A \cup B$.

Claim. $\overline{A} \cap B = \emptyset = A \cap \overline{B}$ (where $\overline{A}$, $\overline{B}$ denote the closure of these sets in (X, d)).

Let $x \in \overline{A} \cap B$. Since B is open in (Y, d), there exists an open set V in (X, d) such that $B = V \cap Y$. Now $x \in B \Rightarrow x \in V$. Since $x \in \overline{A}$, $A \cap V \neq \emptyset$. But $A \cap B = A \cap (V \cap Y) = (A \cap Y) \cap V = A \cap V \Rightarrow A \cap V = \emptyset$. We arrive at a contradiction. Hence $\overline{A} \cap B = \emptyset$. Similarly, $A \cap \overline{B} = \emptyset$. Thus, A and B are separated in (X, d).

(b) $\Rightarrow$ (c): Suppose there exist non-empty separated sets A and B in (X, d) such that $Y = A \cup B$. Since A and B are separated in (X, d), by Lemma 7.2.1 there exist disjoint open sets U and V in (X, d) such that $A \subseteq U$ and $B \subseteq V$. Hence $Y = A \cup B \subseteq U \cup V$, $\emptyset \neq A = A \cap Y \subseteq U \cap Y$ and $\emptyset \neq B = B \cap Y \subseteq V \cap Y$.

(c) $\Rightarrow$ (a): We have $Y = Y \cap (U \cup V) = (Y \cap U) \cup (Y \cap V)$. Also, U is open in $(X, d) \Rightarrow Y \cap U$ is open in (Y, d). Similarly, $Y \cap V$ is open in (Y, d). Further, $U \cap Y \neq \emptyset \neq V \cap Y$. Hence (Y, d) is disconnected. $\square$

Let us end this section with a nice remark. Recall the powerful characterizations of completeness and compactness in terms of sequences. But surprisingly, we do not have sequential characterization of connectedness.

Exercises

7.2.1. Let A be a non-empty connected subset of (X, d). If A is both open and closed in (X, d), then show that A is a component of X.

7.2.2. What are the components of $\mathbb{Q}$? Are the components open?

7.2.3. Let (X, d) be a metric space. Define a relation $\sim$ on X as follows:

$$x \sim y \iff \text{there is a connected subset of } X$$
$$\text{which contains both } x \text{ and } y.$$

Show that $\sim$ is an equivalence relation. Moreover, verify that the equivalence classes are precisely the components of X.

7.2.4. It is known that every singleton set in any metric space is connected. But now think about those metric spaces in which the only connected subsets are singleton. Let us call such metric spaces to be totally disconnected. In other words, a metric space (X, d) is said to be *totally disconnected* if the connected components in X are all singleton sets. Clearly, a metric space in which every point is isolated is totally disconnected but can you think of an example of a totally disconnected metric space which contains a non-isolated point.

7.2.5 (Local version of connectedness). A metric space (X, d) is said to be *locally connected* if for every $x \in X$ and every neighbourhood U of x, there is a connected neighbourhood G of x such that $G \subseteq U$.

(a) Give an example of a connected metric space which is not locally connected.
(b) Give an example of a locally connected metric space which is not connected.
(c) Show that every component of a locally connected space is open.
(d) Let $f : (X, d) \to (Y, \rho)$ be an onto, continuous and open mapping. Then prove that (Y, ρ) is locally connected if (X, d) is so.

Note that locally compact spaces can be equivalently defined as those spaces in which for every point $x \in X$ and every neighbourhood U of x, there is a neighbourhood G of x such that $\overline{G} \subseteq U$ and $\overline{G}$ is compact (see Theorem 5.1.1). But from (a), we get the existence of a metric space (X, d) in which every point has a connected neighbourhood but (X, d) is not locally connected.

7.2.6. Let U be an open subset of $\mathbb{R}^n$. Then show that every component of U is open and there are countably many components.

7.3 Path Connectedness

Though the notion of connectedness is very important in analysis and topology, its definition is somewhat negative in nature. The definition of connectedness ask for non-existence of a certain kind of splitting of the space. A positive approach towards the same sort of problem is provided by path connectedness in which it is feasible to reach any point in the space from any other point along a connected path. This approach is specially useful in analysis and algebraic topology.

The word path is used in a variety of ways in different areas of mathematics. We shall define a path to be any continuous function on $[0, 1]$. The precise definition follows.

Definition 7.3.1. Let (X, d) be a metric space. A continuous function $f : ([0, 1], |\cdot|) \to (X, d)$ is called a *path* (or an *arc*) in X from the point $f(0)$ to the point $f(1)$. The points $f(0)$ and $f(1)$ are called the end points of the path.

Example 7.3.1. Consider the function $f : [0, 1] \to \mathbb{R}^2$ defined as $f(t) = (\cos(2\pi t), \sin(2\pi t))$. Then f represents a path in $\mathbb{R}^2$ from $(1, 0)$ to $(1, 0)$ itself that traces the circle in anti-clockwise direction.

Note that since a path is defined on the compact and connected metric space $([0, 1], |\cdot|)$, it is uniformly continuous and its image is compact as well as connected.

Definition 7.3.2. A metric space (X, d) is called **path connected** (or **arcwise connected**) if for each $x, y \in X$, there is a path in X with end points x and y, that is, there exists a continuous function $f : ([0, 1], |\cdot|) \to (X, d)$ such that $f(0) = x$ and $f(1) = y$.

If $\emptyset \neq Y \subseteq X$ and (Y, d) is path connected, then Y is called path connected in (X, d).

Example 7.3.2. The Euclidean space $\mathbb{R}^n$ is path connected since for any pair of points x, y in $\mathbb{R}^n$, the continuous function $f : [0, 1] \to \mathbb{R}^n$ defined by $f(t) = ty + (1 - t)x$ gives a path from x to y. In fact, this also shows that every normed linear space is pathwise connected.

Note that every open ball $B(x,r)$ in $\mathbb{R}^n$ is also path connected since it can be easily verified that given y, z in $B(x,r)$, $\{ty+(1-t)z : 0 \leq t \leq 1\} \subseteq B(x,r)$. Also see Exercise 7.3.2.

Theorem 7.3.1. *Every path connected metric space (X,d) is connected.*

Proof. Pick $a \in X$ and keep it fixed. Let $x \in X$. Since (X,d) is path connected, there exists a continuous function $f_x : [0,1] \to (X,d)$ such that $f_x(0) = a$ and $f_x(1) = x$. Since $[0,1]$ is connected, $f_x([0,1])$ is connected in (X,d) by Corollary 7.1.1. Now $X = \bigcup \{f_x([0,1]) : x \in X\}$ and $a \in f_x([0,1])$ for all $x \in X$. Hence by Corollary 7.2.1, (X,d) is connected. $\square$

The converse of the previous result may not be true, that is, a connected metric space need not be path connected as seen in the following standard example.

Example 7.3.3 (Topologist's sine curve). Let $A = \{(0,y) \in \mathbb{R}^2 : -1 \leq y \leq 1\}$ and $B = \{(x, \sin\frac{1}{x}) \in \mathbb{R}^2 : 0 < x \leq 1\}$. Suppose $X = A \cup B$ (see Figure 7.1) and consider the Euclidean distance on X

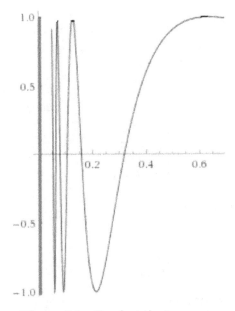

Figure 7.1 Topologist's sine curve.

which is inherited from $\mathbb{R}^2$. Define $f : (0,1] \to B$ by $f(x) = (x, \sin\frac{1}{x})$. Since f is a continuous surjection and $((0,1], | \cdot |)$ is connected, B is connected. Hence by Corollary 7.1.2, the closure of B in $\mathbb{R}^2$ is also connected. But this closure is X itself. Hence X is connected. But it can be shown that no point of A can be connected by a path to any point of B (since the proof is not straightforward, we have divided it into steps in Exercise 7.3.7). Hence X is not path connected.

The previous example also exhibits that the closure of a connected set is connected but the same is not true for path connected sets. Next we show that like connectedness, path connectedness is also preserved by continuous functions.

Theorem 7.3.2. *Let $f : (X,d) \to (Y,\rho)$ be a continuous surjection between two metric spaces. If (X,d) is path connected, then (Y,ρ) is also path connected.*

Proof. Given y_1, y_2 in Y, pick x_1, x_2 in X such that $f(x_1) = y_1$ and $f(x_2) = y_2$. Since (X,d) is path connected, there exists a continuous function $g : [0,1] \to (X,d)$ such that $g(0) = x_1$ and $g(1) = x_2$. Then $f \circ g : [0,1] \to (Y,\rho)$ gives a path from y_1 to y_2. Hence (Y,ρ) is also path connected. $\square$

Now we would like to see under what conditions, the union of a collection of path connected sets is path connected. But for proving this, we need the following proposition.

Proposition 7.3.1. *Let (X,d) be a metric space and x, y, z be in X. If f is a path from x to y and g is a path from y to z in X, then*

$$
h(t) = \begin{cases} f(2t) : & 0 \leq t \leq \dfrac{1}{2} \\ g(2t-1) : & \dfrac{1}{2} \leq t \leq 1 \end{cases}
$$

is a path in X from x to z.

Proof. Since $f(1) = y = g(0)$, $f(2t) = g(2t-1)$ at $t = \frac{1}{2}$. Hence the function h is well-defined. Moreover, h is continuous on $[0,1]$ by pasting lemma (Theorem 2.2.3). $\square$

The path h defined in the preceding proposition is denoted by $g \bullet f$.

Theorem 7.3.3. *Let (X, d) be a metric space.*

(a) *If $\{E_i : i \in I\}$ is a collection of path connected subsets of X such that $E_i \cap E_j \neq \emptyset \,\forall\, i, j \in I$, then $E = \bigcup_{i \in I} E_i$ is path connected in (X, d).*

(b) *If $\{E_i : i \in I\}$ is a collection of path connected subsets of X such that $\bigcap_{i \in I} E_i \neq \emptyset$, then $E = \bigcup_{i \in I} E_i$ is path connected in (X, d).*

(c) *If $\{E_n : n \in \mathbb{Z}\}$ is a sequence of path connected subsets of X such that $E_n \cap E_{n+1} \neq \emptyset \,\forall\, n \in \mathbb{Z}$, then $E = \bigcup_{n \in \mathbb{Z}} E_n$ is also path connected.*

Proof. (a) Let $x, y \in E$. Then suppose $x \in E_i$, $y \in E_j$. Fix any point z in $E_i \cap E_j$. Let $f : [0, 1] \to E_i$ be a path from x to z and let $g : [0, 1] \to E_j$ be a path from z to y. Then $g \bullet f$ is the required path in E from x to y.

(b) This immediately follows from (a).

(c) Let $x, y \in \bigcup\{E_n : n \in \mathbb{Z}\}$. Then there are integers m, n such that $x \in E_m$ and $y \in E_n$. Assume that $m < n$. By part (a), $E_m \cup E_{m+1}$ is path connected. Again (a) implies that $E_m \cup E_{m+1} \cup E_{m+2}$ is path connected. Continuing like this, we get that $E_m \cup \ldots \cup E_n$ is path connected. Thus there is a path in $E_m \cup \ldots \cup E_n \subseteq \bigcup\{E_n : n \in \mathbb{Z}\}$ from x to y. Since x and y were arbitrarily chosen, $\bigcup\{E_n : n \in \mathbb{Z}\}$ is path connected. $\square$

Exercises

7.3.1. Let (X, d) be a metric space and $x, y \in X$. Then show that the following statements are equivalent:

(i) there exists a continuous function $f : [0, 1] \to X$ such that $f(0) = x$ and $f(1) = y$.

(ii) there exists a continuous function $g : [a, b] \to X$ such that $g(a) = x$ and $g(b) = y$.

Thus, to say that the points $x, y \in X$ can be joined by a path, we do not necessarily require a continuous function on $[0, 1]$.

7.3.2. Give an example to show that a ball in a pathwise connected metric space need not be connected.

7.3.3. Let (X, d) be a metric space in which for every $x \in X$ and every neighbourhood U_x of x, there is a path connected ball $B(x, r)$ which is contained in U_x (note that such metric spaces are called *locally path connected*). Then show that every open connected subset of (X, d) is path connected. In particular, we have every open connected subset of $\mathbb{R}^n$ is path connected. But this is not true for closed connected subset of $\mathbb{R}^n$ (consider Topologist's sine curve).

7.3.4. Prove that the ellipse $\{(x, y) \in \mathbb{R}^2 : \frac{x^2}{a^2} + \frac{y^2}{b^2} = 1\}$ is path connected.

7.3.5. Let $\{(X_i, d_i) : 1 \leq i \leq n\}$ be a finite collection of metric spaces. Show that the Cartesian product $\prod_{i=1}^{n} X_i$ equipped with any of the metrics given in Example 1.3.4 is connected (path connected) if and only if (X_i, d_i) is connected (path connected) for all $1 \leq i \leq n$.

7.3.6. Let $\{(X_i, d_i) : i \in \mathbb{N}\}$ be a countable collection of metric spaces. Show that the Cartesian product $\prod_{i=1}^{\infty} X_i$ equipped with the metric given in Example 1.3.10 is connected (path connected) if and only if (X_i, d_i) is connected (path connected) for all $i \in \mathbb{N}$.

7.3.7. Prove the following assertions in order to show that X given in Example 7.3.3 is not path connected.

(a) Suppose $f : [0, 1] \to X$ is a path from $(0, 0)$ to $(1, \sin(1))$. Then $f^{-1}(A)$ is closed.
(b) Let $t_o \in [0, 1]$ be the largest element of $f^{-1}(A)$. Then $t_o < 1$.
(c) The function $f : [t_o, 1] \to X$ given by $f(t) = (x(t), y(t))$ is a path such that $f(t_o) \in A$ and $f((t_o, 1]) \subseteq B$.
(d) Now $x(t_o) = 0$, and for $t \in (t_o, 1]$ we have $x(t) > 0$ and $y(t) = \sin(\frac{1}{x(t)})$.
(e) Prove that f is not continuous at t_o. This gives a contradiction.

7.4　Polygonal Connectedness

Now we move towards defining another stronger form of connectedness where we look for a more specific path between any two points. In particular, we ask for piecewise linear path (and hence it is referred to as polygonal connectedness). Of course, for defining such notion the ambient space needs to possess a linear structure as well.

Definition 7.4.1. Let X be a vector space over $\mathbb{R}$ and A be a subset of X. Then A is said to be *polygonally connected* if every a, $b \in A$ can be joined by a polygonal path in A, that is, there exist some points $a_1, a_2, \ldots, a_{n-1}$ in A such that the line segments $\{(1-\alpha)a_i + \alpha a_{i+1} : \alpha \in [0,1]\}$ are contained in A for every $i \in \{0, 1, \ldots, n-1\}$, where $a_o = a$ and $a_n = b$.

The reader should note that for talking about polygonal connectedness, the ambient space is required to be a vector space, whereas for connectedness and path connectedness it needs to be a metric space (or a topological space in general). Hence if we want to compare all the three forms of connectedness on X then X is required to have a vector space structure as well as a metric space structure (and of course they should be compatible). For that reason, we usually consider a normed linear space.

Example 7.4.1. If S_1 and S_2 are two intersecting balls in $\mathbb{R}^2$, then their union is polygonally connected. In fact, we can also generalize this as follows: if for all $n \in \mathbb{N}$ A_n is a convex set in a vector space X such that $A_i \cap A_{i+1} \neq \emptyset$ for all $i \in \mathbb{N}$ then $\bigcup_{n \in \mathbb{N}} A_n$ is polygonally connected. (Refer to Exercise 1.5.15.)

It is evident that in a normed linear space, every polygonally connected set is path connected and hence connected. But the converse is not true in general. For example, consider a unit circle in $\mathbb{R}^2$ which is path connected but not polygonally connected. Recall that an open connected subset of $\mathbb{R}^n$ is path connected. Interestingly, now we see that it is not just path connected but it is polygonally connected.

Theorem 7.4.1. *Let U be a non-empty open connected subset of a normed linear space $(X, \|\cdot\|)$. Then U is polygonally connected.*

Proof. Fix $x_o \in U$. Consider $U_1 = \{x \in U : \exists$ a polygonal path in U joining x and $x_o\}$. Then $x_o \in U_1$. Now we show that U_1 is both closed and open in U. We start with closedness. Let (x_n)

be a sequence in U_1 which converges to x in U. Since U is open, there exists a $\delta > 0$ such that $B(x, \delta) \subseteq U$. Moreover, for some $n_o \in \mathbb{N}$, $x_n \in B(x, \delta)$ $\forall\, n \geq n_o$. Now $x_{n_o} \in U_1$, thus x_{n_o} and x_o can be joined by a polygonal path in U. Since $B(x, \delta)$ is a convex set in the normed linear space X, x_{n_o} and x can be joined by a line segment in U and hence x and x_o are connected by a polygonal path in U. Consequently, $x \in U_1$. Hence U_1 is closed in U by Corollary 1.5.3.

Now we show that U_1 is open in U. Let $y \in U_1$. Since U is open, there exists an $\epsilon > 0$ such that $B(y, \epsilon) \subseteq U$. Now again by convexity of $B(y, \epsilon)$, we can prove that every point in $B(y, \epsilon)$ can be joined with x_o by some polygonal path. This implies that $B(y, \epsilon) \subseteq U_1$. Hence we have proved that U_1 is non-empty, open as well as closed in a connected set U and hence $U_1 = U$. This means that every point in U can be joined with x_o by some polygonal path. Hence we are done. $\qquad\square$

Remark. As a consequence of the previous theorem, it can be said that all the three forms of connectedness coincides for an open subset of a normed linear space and in particular for open subsets of $\mathbb{R}^n$.

Now the next result gives an application of Theorem 7.4.1 in calculus of several variables.

Theorem 7.4.2. *Let U be an open connected subset of $\mathbb{R}^n$ ($n \geq 2$) and let $f : U \to \mathbb{R}$ be a differentiable function with all partial derivatives 0. Then f is constant on U.*

Proof. Let a, $b \in U$. We prove that $f(a) = f(b)$. By Theorem 7.4.1, there exist some points $x_1, x_2, \ldots, x_{n-1}$ in U such that the line segments $\{(1-t)x_i + tx_{i+1} : t \in [0,1]\}$ are contained in U for every $i \in \{0, 1, \ldots, n-1\}$, where $x_o = a$ and $x_n = b$. We show that f is constant along the path $\gamma : [0,1] \to U : \gamma(t) = (1-t)a + tx_1$.

Let $\gamma(t) = (\gamma_1(t), \ldots, \gamma_n(t))$. Consider the function $g : [0,1] \to \mathbb{R}$ defined as $g(t) = f(\gamma(t)) = f(\gamma_1(t), \ldots, \gamma_n(t))$. Then by chain rule the derivative of g is given by $g'(t) = \frac{\partial f}{\partial \gamma_1}\frac{d\gamma_1}{dt} + \cdots + \frac{\partial f}{\partial \gamma_n}\frac{d\gamma_n}{dt}$. Since the partial derivatives of f are 0, $g'(t) = 0$ $\forall\, t \in [0,1]$. Now by mean value theorem, $g(t) = g(0) = f(a)$ for all $t \in (0,1]$. In particular,

$f(a) = g(1) = f(\gamma(1)) = f(x_1)$. Similarly, we have $f(x_i) = f(x_{i+1})$ for all $i \in \{1, \ldots, n-1\}$. Hence $f(a) = f(b)$. $\qquad\qquad\square$

Remark. If $f : U \to \mathbb{C}$ is a complex analytic function on an open connected subset U of $\mathbb{C}$ with $f'(z) = 0 \;\forall\; z \in U$, then it can be similarly proved that f is constant on U.

Exercise

7.4.1. Show that $\mathbb{R}^2 \setminus A$ is polygonally connected if A is a countable subset of $\mathbb{R}^2$.

Hints to Selected Exercises

Chapter 1

1.3.1 For triangle's inequality, use: $(u+v)^\alpha \leq u^\alpha + v^\alpha$ for all $u \geq 0$ and $v \geq 0$. The function ρ need not be a metric: consider the usual metric on $\mathbb{R}$.

1.3.2 The minimum of two metrics need not be a metric. Take X to be some finite set and try to construct two metrics d_1 and d_2 as in Example 1.3.6.

1.3.3 Define a metric d on X as follows:

$$d(x_1, x_2) = d(x_2, x_1)$$
$$= \left\{ \begin{array}{ll} \rho(x_1, x_2) & : x_1, \ x_2 \in Y \\ \sigma(x_1, x_2) & : x_1, \ x_2 \in X \setminus Y \\ d(x_1, a) + 1 + d(b, x_2) & : x_1 \in Y, \ x_2 \in X \setminus Y \end{array} \right\},$$

where a and b are fixed points in Y and $X \setminus Y$ respectively, and σ is any metric on $X \setminus Y \times X \setminus Y$.

1.3.4 For $x \in \mathbb{R}$, take $f(x) = \tan^{-1}(x)$, $f(+\infty) = \frac{\pi}{2}$ and $f(-\infty) = -\frac{\pi}{2}$.

1.3.7 (i) For triangle inequality: if either $\rho(x, z)$ or $\rho(z, y)$ is 1, $\rho(x, y) \leq \rho(x, z) + \rho(z, y)$ because $\rho(x, y) \leq 1$. Now suppose that both $\rho(x, z)$ and $\rho(z, y)$ are less than 1. Now $\rho(x, y) \leq d(x, y) \leq d(x, z) + d(z, y) = \rho(x, z) + \rho(z, y)$.

(ii) Note that $f(t) = t/(1+t)$ is increasing for $t \geq 0$. Now

$$\delta(x, z) + \delta(z, y) \geq \frac{d(x, z)}{1 + d(x, z) + d(z, y)}$$

227

$$+ \frac{d(z,y)}{1 + d(x,z) + d(z,y)}$$

$$\geq \frac{d(x,z) + d(z,y)}{1 + d(x,z) + d(z,y)}$$

$$\geq \frac{d(x,y)}{1 + d(x,y)} = \delta(x,y).$$

1.3.9 For triangle's inequality, note that

$$1 + d(x,z) \leq 1 + d(x,y) + d(y,z) \leq (1 + d(x,y))(1 + d(y,z)).$$

Now use that $\ln(x)$ is an increasing function.

1.4.3 For counter examples, consider the following subsets of $\mathbb{R}$: $\{(-\frac{1}{n}, \frac{1}{n}) : n \in \mathbb{N}\}$ for (a) and $\{(\frac{1}{n}, n) : n \in \mathbb{N}\}$ for (c).

1.4.4 Consider $A = [0,1)$ and $B = (0,1]$ in $\mathbb{R}$.

1.4.13 $C_b(\mathbb{R})$ is not separable because the sub-collection A of the piecewise linear functions such that $f(n)$ is either 0 or 1, for $n \in \mathbb{N}$, is uncountable. Moreover, $D(f,g) \geq 1$ for $f,\ g \in A$, $f \neq g$.

1.4.14 Let $D = \{x_n : n \in \mathbb{N}\}$ be a countable dense subset of X and Y be a subset of X. For $n,\ m \in \mathbb{N}$, choose $y_{n,m} \in B(x_n, \frac{1}{m}) \cap Y$ if this intersection is non-empty. Then verify that the collection $\{y_{n,m} : n,\ m \in \mathbb{N}\} \subseteq Y$ is dense in Y.

1.4.16 Let D_i be a countable dense subset of X_i for $i \in \mathbb{N}$. Fix $a_i \in D_i$ for each i. Then the set $D = \bigcup_{n \in \mathbb{N}} \{(x_1, \ldots, x_n, a_{n+1}, a_{n+2}, \ldots) : x_i \in D_i \ \forall\ 1 \leq i \leq n\}$ is a countable dense subset of the product space.

1.5.1 We only need to prove that $C(x,r) \subseteq \overline{B(x,r)}$. Let $a \in C(x,r)$. For each $n \in \mathbb{N}$, let us take $a_n = \frac{1}{n}x + (1 - \frac{1}{n})a$. Note that a_n lies on the line segment joining a and x. It is easy to see that $a_n \in B(x,r)$ and $a_n \to a$ as $n \to \infty$. Now use Theorem 1.5.2. For counterexample, think of discrete metric.

1.5.2 (a) Take $A = \{-n : n \in \mathbb{N}\}$ and $B = \{n + \frac{1}{n} : n \in \mathbb{N}\}$. Then both A and B are closed subsets of $\mathbb{R}$. Since $(\frac{1}{n}) \subseteq A + B$ but $0 \notin A + B$, $A + B$ is not closed in $\mathbb{R}$.

1.5.4 (h) $d(x, \bigcap_{i \in I} A_i) \geq \sup\{d(x, A_i) : i \in I\}$. The reverse inequality need not hold true (think of an example).

1.5.11 Take $x = (1, \frac{1}{2}, \frac{1}{3}, \ldots) \in l^\infty$. Construct a sequence in c_{oo} which converges to x. This will show that c_{oo} is not closed in l^∞. Now for showing that it is not open in l^∞, we prove that no open ball around $\mathbf{0} = (0) \in c_{oo}$ is contained in c_{oo}. Suppose, if possible, $B(\mathbf{0}, \frac{1}{n_o}) \subseteq c_{oo}$. Now $(\frac{1}{n_o+1}, \frac{1}{n_o+2}, \ldots) \in B(\mathbf{0}, \frac{1}{n_o})$ but it does not belong to c_{oo}.

1.5.12 Use Corollary 1.5.3 and $|a| \leq |a - b| + |b|$.

Chapter 2

2.1.1 No. Consider the identity map from $(\mathbb{R}, |\cdot|)$ to $(\mathbb{R}, d)$, where d is the discrete metric.

2.1.2 Let $x_n \to x$. Then the sequence $\langle x_1, x, x_2, x, \ldots, x_n, x, \ldots \rangle$ also converges to x. Hence the sequence $\langle f(x_1), f(x), f(x_2), f(x), \ldots, f(x_n), f(x), \ldots \rangle$ converges in (Y, ρ). Since it has a subsequence converging to $f(x)$, the whole sequence converges to the same limit. This implies that $f(x_n) \to f(x)$.

2.1.3 Use Theorems 1.4.4 and 2.1.1.

2.1.4 Use Theorem 2.1.1(e) and Exercise 1.5.4.

2.1.5 Use Corollary 2.1.3.

2.1.7 Let $f : (X, d) \to (Y, \rho)$ be a continuous function and A be a countable subset of X such that $\overline{A} = X$. Then by Theorem 2.1.1, $f(X) = f(\overline{A}) \subseteq \overline{f(A)}$. Note that $f(A)$ is countable. Therefore, $f(X)$ is separable.

2.1.8 Let A be bounded in (X, d). Suppose $f(A)$ is not bounded. Then for every $n \in \mathbb{N}$, there exist $a_n, a'_n \in A$ such that $\rho(f(a_n), f(a'_n)) > n$. By the given condition, (a_n) has a convergent subsequence say (a_{n_k}). Similarly, (a'_{n_k}) has a convergent subsequence which we denote by (a'_{n_k}) itself for convenience. So $a_{n_k} \to a$ and $a'_{n_k} \to a'$. By continuity of f, $f(a_{n_k}) \to f(a)$ and $f(a'_{n_k}) \to f(a')$. $n_k < \rho(f(a_{n_k}), f(a'_{n_k})) \leq \rho(f(a_{n_k}), f(a)) + \rho(f(a), f(a')) + \rho(f(a'), f(a'_{n_k})) < 1 + \rho(f(a), f(a'))$ for $k \geq k_o$ for some $k_o \in \mathbb{N}$. This implies that the increasing sequence (n_k) of naturals is bounded by a fixed number eventually. A contradiction!

2.1.10 (b) Using Theorems 1.5.2 and 2.1.1(d), one can show that every continuous function has a closed graph. We only need to prove that if f is sub-continuous and has closed graph then it is continuous. Suppose, if possible, f is not continuous at $x \in X$. Then there exists an $\epsilon_o > 0$ such that for all $n \in \mathbb{N}$ there exist $x_n \in X$ with $d(x_n, x) < 1/n$ but $\rho(f(x_n), f(x)) > \epsilon_o$. Thus, $x_n \to x$. By sub-continuity of f, there exists a convergent subsequence $(f(x_{k_n}))$ of $(f(x_n))$. Suppose $f(x_{k_n}) \to y$ for some $y \in Y$. Then the sequence $\{(x_{k_n}, f(x_{k_n}))\}$ converges to (x, y) in $(X \times Y, \mu_1)$. But $G(f)$ is closed, hence $(x, y) \in G(f)$. This implies that $y = f(x)$. A contradiction!

2.2.1 (b) Let $\{A_i : i \in I\}$ be such collection and let $x \in \overline{\cup A_i}$. Then there exists a nhood U_x of x such that it intersects A_i for only finitely many i, say $i_1, i_2, \ldots, i_n$. Moreover, every nhood of x intersects $\cup A_i$. Then prove by contradiction that every nhood of x intersects $\bigcup_{k=1}^{n} A_{i_k}$. Hence $x \in \overline{\bigcup_{k=1}^{n} A_{i_k}} = \bigcup_{k=1}^{n} \overline{A_{i_k}}$. Thus, $x \in \overline{A_{i_k}} = A_{i_k}$ for some k. This implies that $\overline{\cup A_i} \subseteq \cup A_i$. The converse inclusion always holds true.

(c) Prove using (b) and Theorem 2.1.1(f).

(d) (i) Consider the collection $\{A_n : n \in \mathbb{N} \cup \{0\}\}$ of sets in the Euclidean space $\mathbb{R}^2$, where $A_0 = \{(0, y) : y \in [0, 1]\}$ and $A_n = \{(\frac{1}{n}, y) : y \in [0, 1]\}$ for $n \in \mathbb{N}$. For $n \in \mathbb{N}$, define $f_n : A_n \to \mathbb{R}$ as: $f_n((\frac{1}{n}, y)) = \frac{1}{n}$. And let $f_0 : A_0 \to \mathbb{R}$ be defined as: $f_0((0, y)) = 1$. Then observe that the sequence $\langle (\frac{1}{n}, 0) \rangle$ converges to $(0, 0)$, but $\frac{1}{n} \not\to 1$.

(ii) Consider the function $f : \mathbb{R} \to \mathbb{R}$ defined as

$$f(x) = \begin{cases} 0 : x \in (-\infty, 0) \\ 1 : \ x \in [0, \infty) \end{cases}.$$

2.2.2 Let $A = \{x \in X : f(x) \geq g(x)\}$ and $B = \{x \in X : f(x) \leq g(x)\}$. Then $X = A \cup B$. Now use pasting lemma.

2.2.4 Let $x_1, \ldots, x_n \in X$. For $i \in \{1, 2, \ldots, n\}$, by Urysohn's Lemma there exists a continuous function $f_i : (X, d) \to [0, 1]$ such that $f_i(x_i) = 1$ and $f_i(x_j) = 0$ for $j \neq i$. Then the set $\{f_i : 1 \leq i \leq n\}$ is linearly independent in $C(X)$. Thus, $\dim(C(X)) \geq n$.

2.3.1 Take $f(x) = \frac{1}{x}$.

2.3.2 Consider the identity map between $(\mathbb{N}, d)$ and $(\mathbb{N}, \rho)$, where d is the usual distance metric and ρ is defined as $\rho(n, m) = |\frac{1}{n} - \frac{1}{m}|$ for $n, m \in \mathbb{N}$. Since every point of $(\mathbb{N}, d)$ and $(\mathbb{N}, \rho)$ is isolated, any function with domain as $(\mathbb{N}, d)$ or $(\mathbb{N}, \rho)$ is continuous. Note that $(\mathbb{N}, d)$ is complete because Cauchy sequences in $(\mathbb{N}, d)$ are eventually constant, whereas $(\mathbb{N}, \rho)$ is incomplete because the Cauchy sequence (n) is not convergent in $(\mathbb{N}, \rho)$.

2.3.3 (a) Use $f(U) = (f^{-1})^{-1}(U)$.

     (b) Take the projection map $\pi_1 : \mathbb{R}^2 \to \mathbb{R} : \pi_1(x, y) = x$. Note that π_1 is open as well as continuous. But the image of the hyperbola $\{(x, y) : xy = 1\}$ under π_1 is $\mathbb{R} \setminus \{0\}$.

     (c) Consider a constant map.

     (d) Use discrete metric.

2.3.4 Note that $|f(x) - f(x')| = \frac{|x' - x|}{xx'} < \frac{1}{a^2}|x - x'|$ for x, $x' \in (a, \infty)$.

2.3.8 (f) If $(x_n) \asymp_u (x'_n)$ then the sequence $\langle x_1, x'_1, x_2, x'_2, \ldots \rangle$ is Cauchy in (X, d) and hence the sequence $\langle f(x_1), f(x'_1), f(x_2), f(x'_2), \ldots \rangle$ is Cauchy in (Y, ρ). This implies that $(f(x_n)) \asymp_u (f(x'_n))$. Conversely, let (x_n) be Cauchy in (X, d). Then $(x_n) \asymp_u (x_n)$. Hence $(f(x_n)) \asymp_u (f(x_n))$. So $(f(x_n))$ is Cauchy.

2.3.11 Let $\epsilon > 0$. Then there exists a $\delta > 0$ such that $d_Z(g(y), g(y')) < \epsilon$ whenever $d_Y(y, y') < \delta$. Now for this δ, apply the uniform continuity of f to get $\delta_1 > 0$ such that $d_X(x, x') < \delta_1$ implies $d_Y(f(x), f(x')) < \delta$. Consequently, we have $d_X(x, x') < \delta_1 \Rightarrow d_Z(g \circ f(x), g \circ f(x')) < \epsilon$.

2.3.13 No. Take $f(x) = g(x) = x$ for $x \in \mathbb{R}$.

2.3.14 Consider $f(x) = x^2$.

2.3.15 Consider the sequences $(\sqrt{n\pi})$ and $(\sqrt{n\pi + \frac{\pi}{2}})$.

2.4.1 Use $|\, \|x\| - \|x'\| \,| \le \|x - x'\|$.

2.4.2 Let $z_1 = (x_1, y_1)$ and $z_2 = (x_2, y_2)$. Then $|f(z_1) - f(z_2)| = |f(x_1, y_1) - f(x_2, y_2)| = |d(x_1, y_1) - d(x_2, y_2)| \le d(x_1, x_2) + d(y_1, y_2) = \mu_1(z_1, z_2)$. Since $\mu_1(z_1, z_2) \le 2\mu_\infty(z_1, z_2) \le 2\mu_2(z_1, z_2)$, f is Lipschitz with any one of these metrics.

2.4.3 Use that $x^n - y^n = (x - y)(x^{n-1} + x^{n-2}y + x^{n-3}y^2 + \cdots + y^{n-1})$.

2.4.4 $\phi(y) - Md(x, y) \le f(y) - Md(x, y) \le f(x)$ for every $f \in F$. Hence $\phi(y) - \phi(x) \le Md(x, y)$. Now interchange the roles of x and y.

2.4.5 Since $\rho(f(x), f(x')) \le Md(x, x') \le M \operatorname{diam}(A)$ for all x, $x' \in A$, $\operatorname{diam}(f(A)) \le M \operatorname{diam}(A) < \infty$.

2.4.6 Let f and g be the identity functions on $\mathbb{R}$. Then $(fg)(x) = x^2$ for $x \in \mathbb{R}$.

2.4.9 By the definition of infimum, for every $n \in \mathbb{N}$, there exists $a_n \in A$ such that $d(x, a_n) < d(x, A) + 1/n$. $(\star)$ Thus, (a_n) is a bounded real sequence and hence it has a convergent subsequence. Let $a_{n_k} \to a$. Since A is closed, $a \in A$. Now by the continuity of the function $y \mapsto d(x, y)$, $d(x, a_{n_k}) \to d(x, a)$. By

$(\star)$, $d(x, A) \le d(x, a) \le d(x, A) + 1/n$ for all $n \in \mathbb{N}$. Hence $d(x, A) = d(x, a)$.

2.4.10 Let $f(x) = \sqrt{x}$ for $x \in [0, \infty)$. We know that $|\sqrt{x} - \sqrt{y}|^2 = |\sqrt{x} - \sqrt{y}||\sqrt{x} - \sqrt{y}| \le |\sqrt{x} + \sqrt{y}||\sqrt{x} - \sqrt{y}| = |x - y|$. Hence $|\sqrt{x} - \sqrt{y}| \le \sqrt{|x - y|}$ for $x, y \ge 0$. Thus f is uniformly continuous (take $\delta = \epsilon^2$). But f is not Lipschitz on $[0, \infty)$ because $\frac{|\sqrt{x} - 0|}{|x - 0|} \to \infty$ as $x \to 0^+$.

2.4.12 (b) Take $f(x) = \sqrt{x}$ for $x \ge 0$. Note that $|\sqrt{x} - \sqrt{y}| \le \sqrt{|x - y|}$ for $x, y \ge 0$.

(c) Let $\alpha = 1 + k$ where $k > 0$. Then, $|\frac{f(x) - f(y)}{x - y}| \le M|x - y|^k$ for $x, y \in I$ with $x \ne y$. Then $\lim_{y \to x} |\frac{f(x) - f(y)}{x - y}| = 0$ and hence $\lim_{y \to x} \frac{f(x) - f(y)}{x - y} = 0$. Thus, $f'(x) = 0$ for all $x \in I$ and hence f is constant on I.

2.4.13 Let (x_n) and (y_n) be two sequences in X and (λ_n) be a sequence in $\mathbb{K}$. Suppose that $x_n \to x$ and $y_n \to y$ in (X, d), then prove that $x_n + y_n \to x + y$ in (X, d). This will prove that f_1 is continuous by Theorem 2.1.2. Similarly for proving the continuity of f_2, suppose that $\lambda_n \to \lambda$ in $(\mathbb{K}, |\cdot|)$ and $x_n \to x$ in (X, d). Then show that $\lambda_n \cdot x_n \to \lambda \cdot x$ in (X, d).

2.5.2 Suppose $f : [0, 1] \to [a, a + \lambda]$ is an onto isometry, where $\lambda > 0$. Then $|f^{-1}(a + \lambda) - f^{-1}(a)| = |a + \lambda - a| = \lambda$. This implies that $\lambda \le 1$. Now $|1 - 0| = |f(1) - f(0)| \le \lambda$. Hence $\lambda = 1$.

2.5.4 Show that $B \subseteq f(A)$ and $f(A) \subseteq B$.

2.5.5 Let D be a countable dense subset of (X, d). Now show that X is isometric to a subset of l^∞ using the following mapping: $x \mapsto f_x$ where $f_x(y) = d(x, y) - d(y, a)$ for $y \in D$ (where $a \in D$ is fixed). Note that $|f_x(y)| = |d(x, y) - d(y, a)| \le d(x, a)$ for all $y \in D$. Thus, $f_x \in l^\infty$.

2.6.4 Use sequential characterization of continuity.

2.6.7 If U_i is open in (X_i, d_i) for each i, then the set $\prod_{i=1}^n U_i$ is open in $(\prod_{i=1}^n X_i, \mu_\infty)$: let $x = (x_1, \ldots, x_n) \in \prod_{i=1}^n U_i$. Then there exists an $\epsilon > 0$ such that $B_{X_i}(x_i, \epsilon) \subseteq U_i$ for all $i \in \{1, 2, \ldots, n\}$, where $B_{X_i}(x_i, \epsilon)$ denotes the open ball with centre x_i and radius ϵ in (X_i, d_i). Consequently, $B_X(x, \epsilon) = \prod_{i=1}^n B_{X_i}(x_i, \epsilon) \subseteq \prod_{i=1}^n U_i$ where $X = \prod_{i=1}^n X_i$ and $B_X(x, \epsilon)$ denotes the ϵ-ball in (X, μ_∞). Then $\prod_{i=1}^n U_i$ is open and hence any arbitrary union of the sets of this form is also open in the product topology.

Conversely, let U be open in (X, μ_∞). If $x = (x_1, \ldots, x_n) \in U$, then there exists a $\delta_x > 0$ such that $x \in B_X(x, \delta_x) \subseteq U$. Thus, $U = \bigcup_{x \in U} B_X(x, \delta_x) = \bigcup_{x \in U} (\prod_{i=1}^n B_{X_i}(x_i, \delta_x))$.

2.6.8 Consider the sequence (f_n) in $C[0,1]$ defined as: $f_n(x) = x^n$ for $x \in [0,1]$. Then show that as $n \to \infty$ $\rho(f_n, f) \to 0$ but $d(f_n, f) \not\to 0$, where $f(x) = 0$ for all $x \in [0,1]$.

2.6.9 Note that $\mathbb{R}$ is complete but $(0,1)$ is incomplete with respect to the usual metric. Now use Proposition 2.3.1 to get a contradiction.

Chapter 3

3.1.2 Note that the function $f : (\mathbb{R}, d) \to ((-\frac{\pi}{2}, \frac{\pi}{2}), |\cdot|) : f(x) = \tan^{-1} x$ is a bijective isometry. Thus $(\mathbb{R}, d)$ is incomplete as $((-\frac{\pi}{2}, \frac{\pi}{2}), |\cdot|)$ is incomplete.

3.1.4 Let f be defined as in Example 3.1.6. Then $f_n(x) \in B(f_{n_\epsilon}(x), \epsilon)$ $\forall n \geq n_\epsilon$ and $\forall x \in X$. Consequently, $f(x) \in B(f_{n_\epsilon}(x), \epsilon)$ $\forall x \in X$, which implies that $\rho(f(x), f_{n_\epsilon}(x)) < \epsilon$ $\forall x \in X$. Since f_{n_ϵ} is bounded, $f \in B(X, Y)$. Moreover, $D(f, f_{n_\epsilon}) < \epsilon$. Now it is easy to prove that $f_n \to f$ in $B(X, Y)$. Thus, $(B(X, Y), D)$ is complete.

3.1.5 First observe that $(\mathbb{R}, \rho)$ is complete by Exercise 2.6.6 and Proposition 2.6.3. Now the proof is similar to that of Example 3.1.6.

3.1.7 Let (x^n) be a sequence in $\prod_{i=1}^\infty X_i$, where $x^n = (x_1^n, x_2^n, \ldots)$. Then observe that (i) $x^n \to x$, where $x = (x_1, x_2, \ldots)$, in $\prod_{i=1}^\infty X_i$ if and only if $x_i^n \to x_i$ in X_i for each i; (ii) (x^n) is Cauchy in $\prod_{i=1}^\infty X_i$ if and only if (x_i^n) is Cauchy in X_i for each i.

3.2.3 (a) $\Rightarrow$ (b): Suppose f is not uniformly continuous on a totally bounded subset A of X. Then there exists an $\epsilon_o > 0$ such that for all $n \in \mathbb{N}$, there exist $a_n, a_n' \in A$ with $d(a_n, a_n') < \frac{1}{n}$ but $\rho(f(a_n), f(a_n')) \geq \epsilon_o$. Since A is totally bounded, by Theorem 3.2.3 (a_n) has a Cauchy subsequence say (a_{n_k}). By Proposition 2.3.2, (a_{n_k}') is also Cauchy. Now verify that the sequence $\langle a_{n_1}, a_{n_1}', a_{n_2}, a_{n_2}', \ldots \rangle$ is also Cauchy. Since f is Cauchy-continuous, the sequence $\langle f(a_{n_1}), f(a_{n_1}'), f(a_{n_2}), f(a_{n_2}'), \ldots \rangle$ is also Cauchy. A contradiction!

(b) $\Rightarrow$ (a): Let (x_n) be Cauchy in X. Then $A = \{x_n : n \in \mathbb{N}\}$ is totally bounded. Hence f is uniformly continuous (and hence Cauchy continuous) on A. Thus, $(f(x_n))$ is Cauchy.

3.2.4 (a) $\Rightarrow$ (b): Let (x_n) be Cauchy in X. Then the set $A = \{x_n : n \in \mathbb{N}\}$ is totally bounded. Hence $f(A)$ is totally bounded. By Theorem 3.2.3, $(f(x_n))$ has a Cauchy subsequence.

(b) $\Rightarrow$ (a): Let A be totally bounded in X. Suppose $f(A)$ is not totally bounded. Then there exists $\epsilon_o > 0$ and $a_1, a_2 \in A$ such that $f(a_2) \notin B(f(a_1), \epsilon_o)$. Now there exists $a_3 \in A$ such that $f(a_3) \notin B(f(a_1), \epsilon_o) \cup B(f(a_2), \epsilon_o)$. Continuing like this, we will get a sequence (a_n) in A such that $\rho(f(a_i), f(a_j)) \geq \epsilon_o$ for all $i, j \in \mathbb{N}$, $i \neq j$. Since A is totally bounded, (a_n) has a Cauchy subsequence say (a_{n_k}). Then $(f(a_{n_k}))$ has a Cauchy subsequence, which is a contradiction.

3.3.2 Take $x_{n+1} = x_n + (1/n)$ and $x_1 = 0$.

3.3.3 Let (A_n) be a sequence of non-empty closed subsets of $\mathbb{R}$ with $A_{n+1} \subseteq A_n \ \forall \ n \in \mathbb{N}$ and $\lim\limits_{n \to \infty} d(A_n) = 0$. Then A_n is bounded for sufficiently large n and hence by the completeness property of $\mathbb{R}$, A_n has supremum and infimum, say x_n and y_n, respectively. Since A_n is closed, $x_n, y_n \in A_n$. Now $A_{n+1} \subseteq A_n$ implies that $y_n \leq y_{n+1} \leq x_{n+1} \leq x_n$. Since every monotone and bounded sequence converges in $\mathbb{R}$, $x_n \to x$ and $y_n \to y$. Moreover, $x, y \in A_n \ \forall \ n$. But $|x - y| \leq d(A_n) \to 0$. So $x = y$. Hence $\bigcap_{n=1}^{\infty} A_n = \{x\}$.

3.3.4 Consider the series expansion of e^x, $e^x = \sum_{n=0}^{\infty} \frac{x^n}{n!}$. Now take the sequence of partial sums.

3.4.1 Suppose there exists a homeomorphism between the completions $(\widehat{X}_d, d)$ and $(\widehat{X}_\rho, \rho)$ of (X, d) and (X, ρ), respectively, that fixes X. Since every continuous function on a complete space is Cauchy-continuous (Exercise 2.3.9), the metrics d and ρ are Cauchy equivalent on X.

Conversely, suppose that the metrics d and ρ are Cauchy equivalent on X. Then the identity map $id : (X, d) \to (X, \rho)$ is Cauchy-continuous. Let $x \in \widehat{X}_d$. Then there exists a sequence (x_n) in X which converges to x in $(\widehat{X}_d, d)$. This implies that (x_n) is Cauchy in (X, d) and hence in (X, ρ). Thus, the sequence (x_n) converges to some x' in $(\widehat{X}_\rho, \rho)$. Suppose (z_n) is another sequence in X that converges to x in $(\widehat{X}_d, d)$. Then there exists some z in $\widehat{X}_\rho$ such that (z_n) converges to z with respect to the metric ρ. Now, the sequence $\langle x_1, z_1, x_2, z_2, \ldots \rangle$ is convergent to x in $(\widehat{X}_d, d)$ and hence the sequence is convergent in $(\widehat{X}_\rho, \rho)$. So $x' = z$. Consequently, the function $f : (\widehat{X}_d, d) \to (\widehat{X}_\rho, \rho) : f(x) = x'$ is well-defined, that is, the definition of $f(x)$, for $x \in \widehat{X}_d$, is independent of the choice of the sequence in X converging to x with respect to d. Note that f is a continuous extension of id. Moreover, f is a homeomorphism.

3.5.3 Suppose for $\epsilon > 0$, we have $\delta > 0$ such that $\rho(f(a), f(a')) < \epsilon$ whenever $a, a' \in A$ with $d(a, a') < \delta$. Now let $x, x' \in \overline{A}$ such that $d(x, x') < \delta/3$. Then there exist sequences (a_n) and (a'_n) in A which converges to x and x', respectively. Now verify the existence of some $n_o \in \mathbb{N}$ such that $d(a_m, a'_p) < \delta \ \forall \ m, p \geq n_o$. Thus $\rho(f(a_m), f(a'_p)) < \epsilon$. Now use the definition of F along with the triangle inequality to get $\rho(F(x), F(x')) \leq \epsilon$.

3.5.4 Consider $f : (0, \infty) \to \mathbb{R}$ defined as: $f(x) = \sin(\frac{1}{x})$.

3.5.6 Use Theorem 3.5.3. Here observe the significance of Theorem 3.5.3.

3.5.7 Suppose (X, d) is complete. Then use Urysohn's Lemma and Corollary 3.5.2 to get the required function. Conversely, suppose, if possible, (X, d) is not complete. Let (x_n) be a Cauchy sequence of distinct terms which does not converge in X. Hence it has no convergent subsequence. Thus, $A = \{x_{2n} : n \in \mathbb{N}\}$ and $B = \{x_{2n+1} : n \in \mathbb{N}\}$ are disjoint and closed subsets of X. Now use the given condition to get the contradiction.

3.5.8 The proof is analogous to that of Theorem 3.5.4.

3.6.1 Suppose $\mathbb{P} = \bigcup_{n \in \mathbb{N}} F_n$, where F_n's are closed in $\mathbb{R}$. Since $F_n \subseteq \mathbb{P}$, $\text{int} F_n = \emptyset$. Now $\mathbb{Q}$ is a countable union of singletons, hence $\mathbb{R}$ is a countable union of closed and nowhere dense sets. This is a contradiction to Baire category theorem.

3.6.2 Let (U_n) be a sequence of open dense sets in $\mathbb{P}$. Then $U_n = V_n \cap \mathbb{P}$ where V_n is open in $\mathbb{R}$. Since U_n is dense in $\mathbb{P}$ and $\mathbb{P}$ is dense in $\mathbb{R}$, it can be easily shown that V_n is dense in $\mathbb{R}$. Now $\mathbb{R}$ is a Baire space by Baire category theorem. Hence $\bigcap_{n \in \mathbb{N}} U_n = (\bigcap_{n \in \mathbb{N}} V_n) \cap (\bigcap_{q \in \mathbb{Q}} \mathbb{R} \setminus \{q\})$ is dense in $\mathbb{R}$. Now show that it is dense in $\mathbb{P}$.

3.6.4 Suppose $\mathbb{R}$ is countable. Then $\mathbb{R} = \bigcup_{x \in \mathbb{R}} \{x\}$ and each set $\{x\}$ is closed in $\mathbb{R}$. Now use Corollary 3.6.3 to get a contradiction.

3.6.5 The set of natural numbers, $\mathbb{N}$, is nowhere dense in $\mathbb{R}$ but it is not nowhere dense in itself.

3.6.6 Consider $\mathbb{Q} = \bigcup_{q \in \mathbb{Q}} \{q\}$.

3.6.8 Use Exercises 3.6.7 and 3.6.5.

3.7.2 (a) Use Corollary 3.7.2 and Exercise 3.6.1.

(b) Let us take Thomae's function, $f : \mathbb{R} \to \mathbb{R} : f(x) = 0$ if $x \notin \mathbb{Q}$, $f(0) = 1$ and $f(x) = 1/n$ if $x = \frac{m}{n}$ where $m \in \mathbb{Z}$ and $n \in \mathbb{N}$ are coprime. Let $p = m/n$ be such a rational number. Then $f(p) = 1/n$. Let $\epsilon < 1/n$ and $\delta > 0$. Since irrationals are dense in $\mathbb{R}$, there exists $a \notin \mathbb{Q}$ in $(p - \delta, p + \delta)$. Thus, $|f(a) - f(p)| = 1/n > \epsilon$. So f is not continuous at rationals. Now let $c \notin \mathbb{Q}$. If $\epsilon > 0$, then by Archimedean property of $\mathbb{R}$ there exists $n_o \in \mathbb{N}$ such that $1/n_o < \epsilon$. Note that the interval $(c-1, c+1)$ contains only finitely many rationals of the aforementioned form m/n with $n < n_o$. Hence we can choose $0 < \delta < 1$ such that $(c-\delta, c+\delta)$ contains rationals of the form m/n with $n \geq n_o$ only. Consequently, $|f(m/n) - f(c)| = 1/n \leq 1/n_o < \epsilon$. Hence f is continuous at irrationals.

3.7.3 Note that $\overline{A} = \{x \in X : d(x, A) = 0\} = \bigcap_{n=1}^{\infty} \{x \in X : d(x, A) < \frac{1}{n}\}$. Moreover, $f(x) = d(x, A)$ is a continuous function.

3.7.4 Let $\epsilon > 0$ and $x_o \in X$. Then by definition of infimum, there exists an $n_o \in \mathbb{N}$ such that $\omega_{n_o}(f, x_o) < \omega(f, x_o) + \epsilon$. If $x \in X$ with $d(x, x_o) < 1/2n_o$, then $B(x, 1/2n_o) \subseteq B(x_o, 1/n_o)$. Thus, $\omega(f, x) \leq \omega_{2n_o}(f, x) \leq \omega_{n_o}(f, x_o) < \omega(f, x_o) + \epsilon$.

Consider $f : \mathbb{R} \to \mathbb{R}$ defined by: $f(x) = 1$ for $x \geq 0$ and $f(x) = 0$ for $x < 0$. Then verify that the corresponding oscillation function $\omega(f, \cdot)$ is not lower semi-continuous at 0.

Chapter 4

4.1.1 (b) $\Rightarrow$ (a): Suppose, if possible, x_o is an accumulation point of X. Then by Theorem 1.5.2, there exists a sequence (x_n) of distinct points in X which converges to x_o. Now the infinite set $\{x_n, x_o : n \in \mathbb{N}\}$ is compact in X (see Example 4.1.3). This is a contradiction.

4.1.2 Let X be any infinite set equipped with the discrete metric d. Then X is closed and bounded in (X, d), but (X, d) is not compact (recall that any compact subset in (X, d) must be finite).

4.1.3 Yes. Let $\mathcal{A} = \{U_i : i \in I\}$ be an open cover of X. Let U_{i_o} be a non-empty open subset of X for some $i_o \in I$. Then $X \backslash U_{i_o}$ is a proper closed subset of X and hence it is compact. Consequently, there exist $i_1, \ldots, i_n$ in I such that $X \setminus U_{i_o} \subseteq \bigcup_{j=1}^{n} U_{i_j}$. Consequently, $\{U_{i_j} : 0 \leq j \leq n\}$ is a finite subcover for X.

4.1.4 Consider the open cover $\{B(x, 1) : x \in F\}$ of F, where $B(x, 1) = \{y \in l^\infty(\mathbb{R}) : D(y, x) < 1\}$. Now for each $n \in \mathbb{N}$, let $x^{(n)}$ be defined so that $x_n^{(n)} = -1$ and $x_j^{(n)} = 1$ for $j \neq n$. Note that $x^{(n)} \in F$. From this, conclude that $\{B(x, 1) : x \in F\}$ does not have a finite subcover for F.

4.1.7 (i) Note that $\partial A = \overline{A} \cap \overline{\mathbb{R}^n \setminus A} = A \cap \overline{\mathbb{R}^n \setminus A}$, since A, being compact, is closed in $\mathbb{R}^n$. Hence, $\partial A \subseteq A$.

We need to show that $A \subseteq \partial A = \overline{A} - \mathrm{int}(A) = A - \mathrm{int}(A)$. If possible, suppose $z \in A$, but $z \notin \partial A = A - \mathrm{int}(A)$. Therefore $z \in \mathrm{int}\, A$. So there exists $\delta > 0$ such that $B(z, \delta) \subseteq A$. Now draw a ray l starting from z in any direction. Note that l is closed in $\mathbb{R}^n$ and hence $A_1 = A \cap l$ is also closed in $\mathbb{R}^n$. Since $A_1 \subseteq A$ and A is compact, A_1 is also compact. So A_1 is bounded in $\mathbb{R}^n$. Let $m = \sup\{d(a, z) : a \in A_1\}$, where d is the Euclidean metric on $\mathbb{R}^n$. Then there exists a sequence (a_n) in A_1 such that $d(a_n, z) \to m$. Since A_1 is compact, there exists a subsequence (a_{k_n}) of (a_n) such that $a_{k_n} \to a_o$ for some $a_o \in A_1$. Hence $d(a_{k_n}, z) \to d(a_o, z)$. But $d(a_{k_n}, z) \to m$ and so $d(a_o, z) = m$ $(\star)$.

Now we claim that $a_o \in \partial A$. If possible, suppose that $a_o \notin \partial A = A - \mathrm{int}\, A$. Since $a_o \in A_1 \subseteq A$, $a_o \in \mathrm{int}\, A$. So there exists $r > 0$ such that $B(a_o, r) \subseteq A$. Note that $a_o + \frac{r}{2} \frac{a_o - z}{|z - a_o|} \in A \cap l = A_1$. Now

$$\left| a_o + \frac{r}{2} \frac{a_o - z}{|z - a_o|} - z \right| = \left| (a_o - z)\left(\frac{r}{2|z - a_o|} + 1 \right) \right|$$

$$= |a_o - z|\left(\frac{r}{2|z - a_o|} + 1 \right) > |a_o - z|.$$

But it contradicts $(\star)$. Hence $a_o \in \partial A$. Since l was an arbitrary ray, we have a boundary point of A in every direction. This contradicts the fact that ∂A is countable. Therefore, $A = \partial A$.

(ii) Consider the compact subset $[0, 1]$ of $\mathbb{R}$.

4.2.1 Take $\mathbb{N}$ with the metric induced by the usual distance.

4.2.2 By definition of supremum, for each $n \in \mathbb{N}$, there exist $x_n,\ y_n \in E$ such that $\delta - 1/n < d(x_n, y_n) \leq \delta$ $(\star)$. Since E is compact, (x_n) has a convergent subsequence, say $x_{n_k} \to x_o$ where $x_o \in E$. Similarly, there exists a subsequence of (y_{n_k}) which converges to some point of E say y_o. For convenience, we can denote that subsequence by (y_{n_k}) itself. Using Proposition 2.5.1, $d(x_{n_k}, y_{n_k}) \to d(x_o, y_o)$. Now conclude using $(\star)$.

4.2.3 Construct a sequence of functions (f_n) in $C[0, 1]$ in the following manner: for all $n \in \mathbb{N}$, f_n is 0 in $[0, \frac{1}{n+2}]$, 1 at $\frac{1}{n+1}$, 0 in $[\frac{1}{n}, 1]$ and linear in the intervals $[\frac{1}{n+2}, \frac{1}{n+1}]$ and $[\frac{1}{n+1}, \frac{1}{n}]$. Then $D(f_n, f_m) = 1$ for $n \neq m$.

4.2.4 Use sequential characterization of compactness.

4.2.5 Suppose (X_i, d_i) is compact for all $i \in \mathbb{N}$. Let $\psi = (x^n)$ be a sequence in $\prod_{i=1}^{\infty} X_i$, where $x^n = (x_1^n, x_2^n, \ldots)$. If π_i denotes the projection on ith component, then the sequence $\pi_1 o\ \psi$ has a convergent subsequence in X_1. Thus, we can find a subsequence ψ_1 of ψ such that $\pi_1 o\ \psi_1$ converges to $x_1 \in X_1$. Now similarly, we can find a subsequence ψ_2 of ψ_1 such that $\pi_2 o\ \psi_2$ converges to $x_2 \in X_2$. Proceed inductively in this manner. Then finally construct a sequence whose nth term is given by the nth term of the sequence ψ_n. Observe that this is a subsequence of ψ which is convergent to $(x_1, x_2, \ldots)$. Consequently, $\prod_{i=1}^{\infty} X_i$ is compact by Theorem 4.2.4.

4.2.6 Use Theorem 4.2.6 and Exercise 3.2.1.

4.3.1 We will prove that $\overline{A} = A$. Let $x_o \in \overline{A}$. Now the function f defined as $f(x) = -d(x, x_o)$ is continuous on (X, d). Hence its restriction to A attains the maximum on A. Since $x_o \in \overline{A}$, there exists a sequence (a_n) in A such that $a_n \to x_o$. Thus, $f(a_n) \to f(x_o)$. Since $f \leq 0$ and $f(a_n) \to 0$, $\sup\{f(x) : x \in A\} = 0$. Hence $x_o \in A$.

4.3.2 Let $V_n = \{x \in X : f(x) < n\}$. Then by Exercise 2.1.9, $\{V_n : n \in \mathbb{N}\}$ is an open cover of X. Now use the compactness of (X, d) to show that $f(X)$ is bounded above. Consequently, by completeness axiom on $\mathbb{R}$ there exists $\sup\{f(x) : x \in X\}$ in $\mathbb{R}$, call it m. By Theorem 1.2.2, there exists $(x_n) \subseteq X$ such that $f(x_n) \to m$. Since (X, d) is compact, Theorem 4.2.4 implies the existence of a convergent subsequence (x_{k_n}) of (x_n). Let $x_{k_n} \to x_o$. Now prove that $f(x_o) = m$.

4.3.3 Use Theorem 2.1.1(d) and Example 4.1.3.

4.3.4 Let $x_o \in X$. Define a sequence $(x_n) \subseteq X$ as follows: $x_n = f(x_{n-1})$ for $n \in \mathbb{N}$. Since X is compact, (x_n) has a cluster point, say $a \in X$. Hence for $\epsilon > 0$, we can choose m, $k \in \mathbb{N}$ such that $d(x_m, a) < \epsilon$ and $d(x_{m+k}, a) < \epsilon$. Thus, $d(x_o, f(X)) \leq d(x_o, x_k) = d(x_m, x_{m+k}) \leq d(x_m, a) + d(a, x_{m+k}) < 2\epsilon$ for all $\epsilon > 0$. So $x_o \in \overline{f(X)}$ by Exercise 1.5.4. Theorem 4.3.1 implies that $f(X)$ is compact and hence by Proposition 4.1.2, $x_o \in f(X)$. The conclusion may not hold true if we drop the compactness of X: take $X = \mathbb{N}$ and $f(n) = n + 1$.

4.3.5 (a) Clearly, f is continuous. Let us define a function, $g(x) = d(x, f(x))$. Then $|g(x) - g(y)| < 2d(x, y)$ and hence it is continuous. Since X is compact, g attains its minimum at some point, say $x_o \in X$. Suppose $f(x_o) \neq x_o$, then $g(f(x_o)) = d(f(x_o), f(f(x_o))) < d(x_o, f(x_o)) = g(x_o)$. This is a contradiction. Thus, $f(x_o) = x_o$. Uniqueness of fixed point is easy to see.

4.3.6 Use Theorem 4.3.1.

4.3.7 (a) Consider the function f defined on A as: $f(x) = d(x, B)$ for $x \in A$ and use Theorem 4.3.2 and Exercise 1.5.4.

(c) Let $A = \mathbb{N}$ and $B = \{n, \ n - \frac{1}{n} : n \in \mathbb{N}\}$.

4.3.9 Suppose G is closed in $X \times Y$ and f is not continuous at some $x_o \in X$. Then there exist an $\epsilon_o > 0$ such that for all $n \in \mathbb{N}$, we can find $x_n \in X$ such that $d(x_n, x_o) < 1/n$ but $\rho(f(x_n), f(x_o)) \geq \epsilon_o$. Thus, $x_n \to x_o$. Since (Y, ρ) is compact, $(f(x_n))$ has a convergent subsequence. So let $f(x_{n_k}) \to y_o$. Then the sequence $\langle (x_{n_k}, f(x_{n_k})) \rangle$ in G converges to (x_o, y_o). Now use the closeness of G to get a contradiction.

Let (x_n) be a sequence of distinct points in a metric space (X, d) such that $x_n \to x$, $x \in X$. Then consider a function $f : \{x_n, x : n \in \mathbb{N}\} \to \mathbb{N}$, defined by $f(x) = 1$ and $f(x_n) = n$ for $n \in \mathbb{N}$.

4.3.10 Use Proposition 3.2.2 and Theorem 4.3.7.

4.3.11 Let $\epsilon > 0$. Since f is uniformly continuous, there is a $\delta > 0$ such that $|f(x) - f(y)| < \epsilon \le \epsilon + Md(x, y)$ whenever $d(x, y) < \delta$ and $M > 0$. Since f is bounded, there exists $M_1 > 0$ such that $|f(x) - f(y)| \le M_1$ for all x, $y \in X$. Let $M = M_1/\delta$. Then $|f(x) - f(y)| \le M\delta \le Md(x, y) \le Md(x, y) + \epsilon$ whenever $d(x, y) \ge \delta$.

4.4.1 A_m is always closed in $\mathbb{R}^2$. But A_m is bounded in $\mathbb{R}^2$ if and only if m is even. Suppose A_m is bounded, then there exists $M > 0$ such that $|x| \le M$ and $|y| \le M$ for all $(x, y) \in A_m$. If m is odd, then the polynomial $p(x) = x^m + (M + 1)^m - 1$ has a real root, say x_o. Then $(x_o, M + 1) \in A_m$, which is a contradiction.

4.4.2 Use the sequential characterization of compactness in a metric space (Theorem 4.2.4).

4.4.6 No. Consider the identity map from $(\mathbb{N}, |\cdot|)$ to $(\mathbb{N}, d)$ where d is the discrete metric.

Chapter 5

5.1.3 Since $\mathbb{Q}$ is countable, there exists a bijection $f : \mathbb{N} \to \mathbb{Q}$. Since every point of $\mathbb{N}$ is isolated with respect to the usual metric, f is continuous. Note that $\mathbb{N}$ is locally compact, but $\mathbb{Q}$ is not.

5.1.4 Apply Proposition 5.1.3 on projection maps.

5.1.5 For each $x \in K$, we get $\epsilon_x > 0$ such that $\overline{B(x, \epsilon_x)} \subseteq U$ and $\overline{B(x, \epsilon_x)}$ is compact (by Theorem 5.1.1). Since K is compact, $K \subseteq \bigcup_{i=1}^{n} B(x_i, \epsilon_{x_i})$. Now use that $\overline{\bigcup_{i=1}^{n} B(x_i, \epsilon_{x_i})} = \bigcup_{i=1}^{n} \overline{B(x_i, \epsilon_{x_i})}$.

5.1.6 (a) $(0, 1)$ (Corollary 5.1.2).

(b) The proof is analogous to that of Theorem 3.6.2. (Choose C_i which are compact as well. The Cauchy sequence (x_n) lies eventually in the compact spaces C_i.)

(c) Refer to Exercise 3.6.2.

5.2.5 Note that $|I(x) - I(y)| \le 2d(x, y)$.

5.3.3 (a) $\Rightarrow$ (b): Let σ be a metric on X which is equivalent to d. Hence the identity map $I : (X, d) \to (X, \sigma)$ is a homeomorphism. Now suppose that (x_n) is cofinally Cauchy in (X, d). Since (X, d) is cofinally complete, (x_n) has a convergent subsequence, say (x_{n_k}) (Theorem 1.6.1). By the continuity of I, (x_{n_k}) also converges in (X, σ) and hence (x_n) is cofinally Cauchy with respect to the metric σ.

(b) $\Rightarrow$ (a): If (X, d) is not cofinally complete, then there exists a cofinally Cauchy sequence (x_n) of distinct points in (X, d) with no cluster point. Thus the set $A = \{x_n : n \in \mathbb{N}\}$ is closed and discrete and hence the function $f : (A, d) \to \mathbb{R}$ defined by $f(x_n) = n$ is continuous. By Tietze's extension theorem, there exists a real-valued continuous extension $\widetilde{f}$ of f to (X, d). Now define a metric σ on X as follows: $\sigma(x, y) = d(x, y) + |f(x) - f(y)|$ for x, $y \in X$. Then σ is equivalent to d (see Exercise 2.6.2). By (b), (x_n) is cofinally Cauchy in (X, σ). Consequently, the sequence $(f(x_n)) = (n)$ is cofinally Cauchy in $\mathbb{R}$. A contradiction!

5.3.5 (c) $\Rightarrow$ (a): Suppose f is not uniformly continuous. Then there exists an $\epsilon_o > 0$ such that $\forall n \in \mathbb{N}, \exists x_n, z_n \in X$ with $d(x_n, z_n) < \frac{1}{n}$ but $\rho(f(x_n), f(z_n)) > \epsilon_o$. Hence $(x_n) \asymp (z_n)$ but $(f(x_n)) \not\asymp^c (f(z_n))$, a contradiction.

5.4.1 For $\mathbb{R}^n$, use Heine–Borel theorem. For l^p: let A be contained in a ball, $B(x_o, r)$, in l^p, where $x_o \in l^p$ and $r > 0$. Let $\epsilon > 0$. Choose $m \in \mathbb{N}$ such that $\frac{r}{m} < \epsilon$. Let $a \in A$ and define $\xi_i = x_o + \frac{i}{m}(a - x_o)$ for all $i = 1, 2, \dots, m$ and $\xi_o = x_o$. Then one can verify that $d(\xi_{i-1}, \xi_i) < \epsilon \ \forall i = 1, 2, \dots, m$. Hence every element of A can be joined with x_o by an ϵ-chain of length m.

5.4.3 Use Theorem 5.4.2 and Example 2.6.4.

5.4.4 The Bourbaki–Cauchy sequences in X are eventually constant.

Chapter 6

6.1.1 Suppose $f(x) = \lim_{n \to \infty} f_n(x)$ for every $x \in X$. If f is not increasing, then there exist x, $y \in X$ such that $x < y$ but $f(x) > f(y)$. Let $\epsilon_o = f(x) - f(y)$. Then there exists $n_o \in \mathbb{N}$ such that $|f_n(x) - f(x)| < \frac{\epsilon_o}{3}$ and $|f_n(y) - f(y)| < \frac{\epsilon_o}{3} \ \forall n \geq n_o$. Now $f_{n_o}(x) - f_{n_o}(y) = (f_{n_o}(x) - f(x)) + (f(x) - f(y)) + (f(y) - f_{n_o}(y)) > \frac{-\epsilon_o}{3} + \epsilon_o - \frac{\epsilon_o}{3} = \frac{\epsilon_o}{3} > 0$. A contradiction! Note that the same result may not hold true for strictly increasing functions: consider $f_n(x) = x/n$.

6.1.2 Use Theorem 2.1.1(d), Example 4.1.3 and Theorem 6.1.2.

6.1.3 Let $f_n : \mathbb{R} \to \mathbb{R}$ be defined as $f_n(\frac{1}{n}) = \frac{1}{n}$ and otherwise it is 0.

6.1.5 Suppose (f_n) is a Cauchy sequence in $(\text{Lip}(X), \|\cdot\|)$. Let $\epsilon > 0$. Then there exists $n_o \in \mathbb{N}$ such that $|f_m(x_o) - f_n(x_o)| < \epsilon$ and $L(f_m - f_n) < \epsilon \ \forall \ m, n \geq n_o$. Thus, for all $x \in X$ and for all $m, n \geq n_o$, we have

$$|f_m(x) - f_n(x) - f_m(x_o) + f_n(x_o)| < \epsilon \, d(x, x_o).$$

Then $(f_n(x))$ is Cauchy in $\mathbb{R} \ \forall \ x \in X$. Let $f(x) := \lim_{n \to \infty} f_n(x) \ \forall \ x \in X$. Since a Cauchy sequence is bounded, by

Theorem 6.1.4 $f \in \text{Lip}(X)$. Now the claim is $\|f_n - f\| \to 0$: let $\epsilon > 0$. Since (f_n) is Cauchy, for $x, x' \in X$ we have $|f_m(x) - f_n(x) - f_m(x') + f_n(x')| < \epsilon \, d(x, x') \ \forall \ m, n \geq n_o$. Now let $m \to \infty$. Then $|(f - f_n)(x) - (f - f_n)(x')| \leq \epsilon \, d(x, x') \ \forall \ n \geq n_o$.

6.2.2 The sequence of partial sums (s_n) for the series $\sum_{n \in \mathbb{N}} f_n$ is an increasing sequence in $C[a, b]$. Now use Dini's theorem to conclude.

6.2.4 Let $\epsilon > 0$. By Proposition 6.2.1, (f_n) is uniformly Cauchy on D. Thus, there exists $n_o \in \mathbb{N}$ such that $|f_n(y) - f_m(y)| < \epsilon/3 \ \forall \ n, m \geq n_o$ and $\forall \ y \in D$. Let $x \in X$. Then there exists $(y_n) \subseteq D$ such that $y_n \to x$. Now use the continuity of the functions to prove that (f_n) is uniformly Cauchy on X.

6.3.3 (c) Take any bounded continuous function $f : \mathbb{R} \to \mathbb{R}$ which is not uniformly continuous.

     (d) Let $X = \mathbb{N}$ and $S = \{f_n : n \in \mathbb{N}\}$ where $f_n(n) = 1$ and otherwise it is zero.

6.3.9 Suppose for each $n \in \mathbb{N}$, there exists $M_n > 0$ such that $|f_n(x)| < M_n \ \forall \ x \in X$. Since (f_n) converges uniformly to f on X, there exists $n_o \in \mathbb{N}$ such that $|f_n(x) - f(x)| < 1$ for all $n \geq n_o$ and for all $x \in X$. As a consequence, for $n \geq n_o$ and $x \in X$ we have, $|f_n(x)| \leq |f_n(x) - f(x)| + |f(x) - f_{n_o}(x)| + |f_{n_o}(x)| < 2 + M_{n_o}$. Let $M = \max\{M_1, M_2, \ldots, M_{n_o-1}, 2 + M_{n_o}\}$. Then $|f_n(x)| < M \ \forall \ n \in \mathbb{N}$ and $\forall \ x \in X$.

6.3.10 (c): Yes. If $S \subseteq C(X)$ is equicontinuous then so is $\overline{S}$.

6.4.2 Note that the dimension of $\mathbb{R}^2$ over $\mathbb{R}$ is 2.

6.4.3 Follows from Theorem 6.4.5.

6.4.4 (b) Use Theorem 4.3.2.

6.4.5 Note that $\mathrm{Lip}(X)$ contains the constant functions and hence it vanishes nowhere on X. Further, it separates the points of X: let $x_1 \neq x_2 \in X$, then the function $f(x) = d(x, x_1)$ belongs to $\mathrm{Lip}(X)$ and $f(x_1) \neq f(x_2)$. Now use Exercise 6.4.4 and Theorem 6.4.3.

6.4.7 From Corollary 6.4.8, there exists a sequence of polynomials (p_n) which converges uniformly to f on $[0, 1]$. Then $\int_0^1 f^2(x)dx = \lim_{n\to\infty} \int_0^1 f(x)p_n(x)dx = 0$. Now using continuity of f, show by contradiction that $f = 0$ on $[0, 1]$.

6.4.8 Let $A = \{x_n : n \in \mathbb{N}\}$ be a countable dense subset of X and let $f_n(x) = d(x, x_n)$. If A is the subalgebra of $C(X)$ generated by the functions $1, f_1, f_2, \ldots$, then show that $\overline{A} = C(X)$ (use Theorem 6.4.6).

6.4.9 (a) Example can be easily constructed using Weierstrass approximation theorem (Corollary 6.4.8).

    (b) Consider $f_n(x) = x^{n+1}/(n+1)$ for $x \in [0, 1]$. Then $f_n \to 0$ uniformly because $|f_n(x)| \leq 1/(n+1) \to 0$ as $n \to \infty$ for any $x \in [0, 1]$. But $f_n'(x) = x^n$. Now see Example 6.1.2.

    (c) Let $x \in I$ and m, $n \in \mathbb{N}$. Apply Lagrange mean value theorem to the function $f_m - f_n$ on the interval with endpoints x_o and x. Now use Proposition 6.2.1.

6.4.10 Let $A_r = \{f \in A : f(X) \subseteq \mathbb{R}\}$. If $f \in A$, then $\mathrm{Re}f = (f + \overline{f})/2 \in A_r$ and similarly, $\mathrm{Im}f \in A_r$. Apply Theorem 6.4.6 on A_r.

Chapter 7

7.1.1 Take $f(x) = (x+1)/2$.

7.1.2 Let $A = \{(x, y) : (x-1)^2 + y^2 \leq 1\} \cup \{(x, y) : (x+1)^2 + y^2 \leq 1\}$. Then A is connected but A° is not connected.

7.1.3 Consider the function $f(x) = x^m$. Then $f(0) = 0$. Let $x_o = \max\{a, 1\}$. Then $f(x_o) = x_o^m > a$. Thus, $f(0) < a < f(x_o)$. Now use intermediate value theorem.

7.1.4 If possible, suppose x and y are two distinct elements in a countable connected subset A of (X, d). Then there exists $0 <$

$r < d(x,y)$ such that $B(x,r) \cap A = B[x,r] \cap A$. Now use Theorem 7.1.1 to get a contradiction.

7.1.5 Suppose (X,d) is not connected, then there exists a non-constant continuous function $g : (X,d) \to \{0,1\}$, where $\{0,1\}$ is equipped with the discrete metric. Let $h : \{0,1\} \to \mathbb{R}$ be defined by: $h(0) = 0$ and $h(1) = 1$. Then h is continuous and so is $h \circ g$ but $h \circ g$ does not satisfy IVP.

7.1.6 Note that if $f : X \to Y$ is a homeomorphism, then $f : A \to f(A)$ is also a homeomorphism for $A \subseteq X$. Moreover, we know that if we remove a point from a circle then it will still remain connected. So use Corollary 7.1.2. For the second part, use Theorem 4.3.3.

7.1.7 First think about connected subsets of $\mathbb{Q}$. Then look at some characteristic function defined on $\mathbb{Q}$.

7.2.2 Every component of $\mathbb{Q}$ is a singleton.

7.2.5 (a) Topologist's sine curve (Example 7.3.3).

(b) $(0,1) \cup (2,3)$.

(c) Let C be a component in (X,d) and $x \in C$. Then there exists a connected neighbourhood G of x. Since $x \in C \cap G$, $C \cup G$ is connected. But C is a component, hence $C = C \cup G$ which further implies that $G \subseteq C$.

7.2.6 For the first part, use Exercise 7.2.5(c). Let D be a countable dense subset in $\mathbb{R}^n$. Since each component in U is open, each component contains a point of D. Now distinct components are disjoint and D is countable. Hence there are countably many components.

7.3.1 Use the homeomorphism $h : [0,1] \to [a,b]$ defined as $h(x) = xb + (1-x)a$. And the inverse function h^{-1}.

7.3.2 Let $X = \mathbb{R}^2 \setminus \{(0,y) : -1 \le y \le 1\}$ and consider the open unit ball in X with centre $(\frac{1}{2},0)$.

7.3.3 Let U be a non-empty open connected set. Fix $x_o \in U$. Consider $U_1 = \{x \in U : \exists$ a path in U joining x and $x_o\}$. Then $x_o \in U_1$. Now show that U_1 is both closed and open in U. Thus, $U_1 = U$. Also refer to the proof of Theorem 7.4.1.

7.3.4 The ellipse is the image of $[0,2\pi]$ under the continuous mapping $f(t) = (a\cos(t), b\sin(t))$.

7.3.6 For path connectedness, use Exercise 2.2.3, Theorem 7.3.2 and projection maps.

7.4.1 Suppose $\mathbb{R}^2 \setminus A$ is not polygonally connected. Let x, $y \in \mathbb{R}^2 \setminus A$. Consider a straight line l between x and y which is not passing through them. Now draw line segments joining x and points on l. Similarly draw line segments joining y and points on l. This gives uncountable number of polygonal paths between x and y. Since $\mathbb{R}^2 \setminus A$ is not polygonally connected, each such path contains distinct points of A. This implies that A is uncountable.

Bibliography

[1] C. D. Aliprantis and O. Burkinshaw. *Principles of Real Analysis*, 3rd edn. Academic Press, Inc., San Diego, CA, 1998.

[2] M. Atsuji. Uniform continuity of continuous functions of metric spaces. *Pacific J. Math.*, 8:11–16; erratum, 941, 1958.

[3] G. Beer. Metric spaces on which continuous functions are uniformly continuous and Hausdorff distance. *Proc. Am. Math. Soc.*, 95(4):653–658, 1985.

[4] G. Beer. UC spaces revisited. *Am. Math. Monthly*, 95(8):737–739, 1988.

[5] G. Beer. *Topologies on Closed and Closed Convex Sets*, Volume 268 of Mathematics and Its Applications. Kluwer Academic Publishers Group, Dordrecht, 1993.

[6] G. Beer. Between compactness and completeness. *Topol. Appl.*, 155(6):503–514, 2008.

[7] G. Beer. *Bornologies and Lipschitz Analysis*. CRC Press, Boca Raton, 2023.

[8] G. Beer and M. I. Garrido. Bornologies and locally Lipschitz functions. *Bull. Aust. Math. Soc.*, 90(2):257–263, 2014.

[9] B. Bolzano. *Rein analytischer Beweis des Lehrsatzes daß zwischen je zwey Werthen, die ein entgegengesetzetes Resultat gewähren, wenigstens eine reelle Wurzel der Gleichung liege*. Gottlieb Haase, Prag, 1817.

[10] J. Borsík. Mappings that preserve Cauchy sequences. *Časopis Pěst. Mat.*, 113(3):280–285, 1988.

[11] N. Bourbaki. *Elements of Mathematics. General Topology. Part 1*. Addison-Wesley Publishing Co., Paris, 1966.

[12] D. Burton and J. Coleman. Quasi-Cauchy sequences. *Am. Math. Monthly*, 117(4):328–333, 2010.

[13] J. Cabello Sánchez. $U(X)$ as a ring for metric spaces X. *Filomat*, 31(7):1981–1984, 2017.

[14] M. A. Chaves. Spaces where all continuity is uniform. *Am. Math. Monthly*, 92(7):487–489, 1985.

[15] E. T. Copson. *Metric Spaces*, Cambridge Tracts in Mathematics and Mathematical Physics, No. 57. Cambridge University Press, London, 1968.

[16] H. H. Corson. The determination of paracompactness by uniformities. *Am. J. Math.*, 80:185–190, 1958.

[17] R. Doss. On uniformly continuous functions in metrizable spaces. *Proc. Math. Phys. Soc. Egypt*, 3:1–6, 1947.

[18] R. V. Fuller. Relations among continuous and various non-continuous functions. *Pacific J. Math.*, 25:495–509, 1968.

[19] M. I. Garrido and A. S. Meroño. New types of completeness in metric spaces. *Ann. Acad. Sci. Fenn. Math.*, 39(2):733–758, 2014.

[20] C. Goffman and G. Pedrick. *First Course in Functional Analysis*. Prentice-Hall, Inc., New Jersey, 1965.

[21] J. Hejcman. Boundedness in uniform spaces and topological groups. *Czechoslovak Math. J.*, 9(84):544–563, 1959.

[22] E. Hewitt. The rôle of compactness in analysis. *Am. Math. Monthly*, 67:499–516, 1960.

[23] A. Hohti. On uniform paracompactness. *Ann. Acad. Sci. Fenn. Ser. A I Math. Dissertationes*, (36):1–46, 1981.

[24] N. R. Howes. On completeness. *Pacific J. Math.*, 38:431–440, 1971.

[25] H. Hueber. On uniform continuity and compactness in metric spaces. *Am. Math. Monthly*, 88(3):204–205, 1981.

[26] E. Kreyszig. *Introductory Functional Analysis with Applications*, Wiley Classics Library. John Wiley & Sons, Inc., New York, 1989.

[27] S. Kundu, M. Aggarwal, and L. Gupta. *Cofinally Complete Metric Spaces and Related Functions*. World Scientific, Singapore, 2023.

[28] S. Kundu, M. Aggarwal, and S. Hazra. Finitely chainable and totally bounded metric spaces: Equivalent characterizations. *Topol. Appl.*, 216:59–73, 2017.

[29] S. Kundu and T. Jain. Atsuji spaces: Equivalent conditions. *Topol. Proc.*, 30(1):301–325, 2006.

[30] M. Mandelkern. On the uniform continuity of Tietze extensions. *Arch. Math.*, 55(4):387–388, 1990.

[31] G. Marino, G. Lewicki, and P. Pietramala. Finite chainability, locally Lipschitzian and uniformly continuous functions. *Z. Anal. Anwendungen*, 17(4):795–803, 1998.

[32] R. A. McCoy, S. Kundu, and V. Jindal. *Function Spaces with Uniform, Fine and Graph Topologies*, SpringerBriefs in Mathematics. Springer, Cham, 2018.

[33] E. J. McShane. Extension of range of functions. *Bull. Am. Math. Soc.*, 40(12):837–842, 1934.

[34] A. A. Monteiro and M. M. Peixoto. Le nombre de Lebesgue et la continuité uniforme. *Portugal Math.*, 10:105–113, 1951.

[35] J. Nagata. On the uniform topology of bicompactifications. *J. Inst. Polytech. Osaka City Univ. Ser. A. Math.*, 1:28–38, 1950.

[36] V. Niemytzki and A. Tychonoff. Beweis des satzes, dass ein metrisierbarer raum dann und nur dann kompakt ist, wenn er in jeder metrik vollständig ist. *Fund. Math.*, 12:118–120, 1928.

[37] C. G. C. Pitts. *Introduction to Metric Spaces*, University Mathematical Texts. Oliver & Boyd [Longman Group Ltd.], Edinburgh, 1972.

[38] J. Rainwater. Spaces whose finest uniformity is metric. *Pacific J. Math.*, 9:567–570, 1959.

[39] M. D. Rice. A note on uniform paracompactness. *Proc. Am. Math. Soc.*, 62(2):359–362, 1977.

[40] K. A. Ross. *Elementary Analysis: The Theory of Calculus*, 2nd edn. Undergraduate Texts in Mathematics, Springer, New York, 2013.

[41] H. L. Royden and P. M. Fitzpatrick. *Real Analysis*, 4th edn. Prentice Hall, New Jersey, 2010.

[42] W. Rudin. *Principles of Mathematical Analysis*, 3rd edn. International Series in Pure and Applied Mathematics, McGraw-Hill Book Co., New York-Auckland-Düsseldorf, 1976.

[43] M. Ó. Searcóid. *Metric Spaces*. Springer Undergraduate Mathematics Series, Springer-Verlag London, Ltd., London, 2007.

[44] B. Simon. *Real Analysis: A Comprehensive Course in Analysis, Part 1*. American Mathematical Society, USA, 2015.

[45] J. Smith. Review of "A note on uniform paracompactness" by Michael D. Rice. *Math. Rev.*, 55(#9036), 1978.

[46] R. F. Snipes. Functions that preserve Cauchy sequences. *Nieuw Arch. Wisk. (3)*, 25(3):409–422, 1977.

[47] K. R. Stromberg. *An Introduction to Classical Real Analysis*. Wadsworth International, Belmont, California, 1981.

[48] Gh. Toader. On a problem of Nagata. *Mathematica*, 20(43)(1):77–79, 1978.

[49] W. C. Waterhouse. On UC spaces. *Am. Math. Monthly*, 72:634–635, 1965.

[50] S. Willard. *General Topology*. Addison-Wesley Publishing Co., Reading, MA–London–Don Mills, Ontario, 1970.

Appendix

Some Useful Inequalities

Theorem A.1 (Young's inequality). *Let a and b be non-negative real numbers and $0 < \lambda < 1$ be a fixed real number. Then*

$$a^\lambda b^{1-\lambda} \le \lambda a + (1 - \lambda)b.$$

Proof. The inequality certainly holds when either a or b is zero. So let us assume that $a > 0$ and $b > 0$. Since the function $\ln x$ is concave on $(0, \infty)$, $\ln(\lambda a + (1 - \lambda)b) \ge \lambda \ln a + (1 - \lambda) \ln b$. This implies that $a^\lambda b^{1-\lambda} \le \lambda a + (1 - \lambda)b$ because the function e^x is increasing on $\mathbb{R}$. $\qquad\square$

The main purpose of Young's inequality is to prove the next inequality.

Theorem A.2 (Hölder's inequality). *If $p > 1$ and q is such that $\frac{1}{p} + \frac{1}{q} = 1$, then we have*

$$\sum_{i=1}^n |x_i y_i| \le \left(\sum_{i=1}^n |x_i|^p \right)^{1/p} \left(\sum_{i=1}^n |y_i|^q \right)^{1/q}$$

for $(x_1, x_2, \ldots, x_n)$ and $(y_1, y_2, \ldots, y_n)$ in $\mathbb{R}^n$ or $\mathbb{C}^n$.

Proof. If either $\sum_{i=1}^n |x_i|^p$ or $\sum_{i=1}^n |y_i|^q$ is zero, then we are done. Hence we assume that they both are non-zero. For convenience, let us denote $(\sum_{i=1}^n |x_i|^p)^{1/p}$ by X and $(\sum_{i=1}^n |y_i|^q)^{1/q}$ by Y. Now for every

$i \in \{1, 2, \ldots, n\}$ we apply Young's inequality for $a = \frac{|x_i|^p}{X^p}$, $b = \frac{|y_i|^q}{Y^q}$ and $\lambda = 1/p$. Hence we get,

$$\frac{|x_i y_i|}{XY} \leq \frac{1}{p} \frac{|x_i|^p}{X^p} + \frac{1}{q} \frac{|y_i|^q}{Y^q} \quad \forall \, i \in \{1, 2, \ldots, n\}.$$

Now by adding all the n inequalities, we get

$$\frac{\sum_{i=1}^{n} |x_i y_i|}{XY} \leq \frac{1}{p} \frac{\sum_{i=1}^{n} |x_i|^p}{X^p} + \frac{1}{q} \frac{\sum_{i=1}^{n} |y_i|^q}{Y^q} = \frac{1}{p} + \frac{1}{q} = 1. \qquad \square$$

Remark. When we particularly take $p = 2$, then Hölder's inequality reduces to

$$\left(\sum_{i=1}^{n} |x_i y_i| \right)^2 \leq \left(\sum_{i=1}^{n} |x_i|^2 \right) \left(\sum_{i=1}^{n} |y_i|^2 \right).$$

Since this special case is widely applicable especially in functional analysis and probability theory, it is given a special name *Cauchy–Schwarz inequality*.

Now we prove the main inequality which is required to prove the triangle inequality for some significant metrics in Chapter 1.

Theorem A.3 (Minkowski's inequality for finite sums). *For* $p \geq 1$, *we have*

$$\left(\sum_{i=1}^{n} |x_i + y_i|^p \right)^{1/p} \leq \left(\sum_{i=1}^{n} |x_i|^p \right)^{1/p} + \left(\sum_{i=1}^{n} |y_i|^p \right)^{1/p},$$

where $(x_1, x_2, \ldots, x_n)$ *and* $(y_1, y_2, \ldots, y_n)$ *in* $\mathbb{R}^n$ *or* $\mathbb{C}^n$.

Proof. We only need to prove the case for $p > 1$ and $\sum_{i=1}^{n} |x_i + y_i|^p \neq 0$.

$$\sum_{i=1}^{n} |x_i + y_i|^p \leq \sum_{i=1}^{n} |x_i||x_i + y_i|^{p-1} + \sum_{i=1}^{n} |y_i||x_i + y_i|^{p-1}$$

$$\leq \left(\sum_{i=1}^{n}|x_i|^p\right)^{1/p}\left(\sum_{i=1}^{n}|x_i+y_i|^{(p-1)q}\right)^{1/q}$$

$$+\left(\sum_{i=1}^{n}|y_i|^p\right)^{1/p}\left(\sum_{i=1}^{n}|x_i+y_i|^{(p-1)q}\right)^{1/q},$$

(by Hölder's inequality)

$$=\left[\left(\sum_{i=1}^{n}|x_i|^p\right)^{1/p}+\left(\sum_{i=1}^{n}|y_i|^p\right)^{1/p}\right]\left(\sum_{i=1}^{n}|x_i+y_i|^p\right)^{1/q}.$$

Now by dividing both sides by $(\sum_{i=1}^{n}|x_i+y_i|^p)^{1/q}$, we get the required result. □

Theorem A.4 (Minkowski's inequality for infinite sums).
Suppose $p \geq 1$, *and* (x_n) *and* (y_n) *are sequences in* $\mathbb{R}$ *or* $\mathbb{C}$ *such that* $\sum_{i=1}^{\infty}|x_i|^p$ *and* $\sum_{i=1}^{\infty}|y_i|^p$ *are convergent. Then the series* $\sum_{i=1}^{\infty}|x_i+y_i|^p$ *is convergent. Moreover we have,*

$$\left(\sum_{i=1}^{\infty}|x_i+y_i|^p\right)^{1/p}\leq\left(\sum_{i=1}^{\infty}|x_i|^p\right)^{1/p}+\left(\sum_{i=1}^{\infty}|y_i|^p\right)^{1/p}.$$

Proof. By Theorem A.3, for any $n \in \mathbb{N}$ we have

$$\left(\sum_{i=1}^{n}|x_i+y_i|^p\right)^{1/p}\leq\left(\sum_{i=1}^{n}|x_i|^p\right)^{1/p}+\left(\sum_{i=1}^{n}|y_i|^p\right)^{1/p}$$

$$\leq\left(\sum_{i=1}^{\infty}|x_i|^p\right)^{1/p}+\left(\sum_{i=1}^{\infty}|y_i|^p\right)^{1/p}.$$

Thus, $\left\{(\sum_{i=1}^{n}|x_i+y_i|^p)^{1/p}\right\}_{n\geq 1}$ is a monotonically increasing sequence of real numbers which is bounded above by $(\sum_{i=1}^{\infty}|x_i|^p)^{1/p}+$

$(\sum_{i=1}^{\infty} |y_i|^p)^{1/p} < \infty$. Consequently, the series $\sum_{i=1}^{\infty} |x_i + y_i|^p$ is convergent and

$$\left(\sum_{i=1}^{\infty} |x_i + y_i|^p\right)^{1/p} \leq \left(\sum_{i=1}^{\infty} |x_i|^p\right)^{1/p} + \left(\sum_{i=1}^{\infty} |y_i|^p\right)^{1/p}. \qquad \square$$

Index

Printed in the USA
CPSIA information can be obtained
at www.ICGtesting.com
JSHW011931241023
50596JS00003B/37

9 789811 278914